1988 Colloquium Reader

Technology, Security, and Arms Control for the 1990s

Edited by

Elizabeth J. Kirk

Program on Science, Arms Control,
and National Security of the
American Association for the Advancement of Science

These essays were prepared for the AAAS Colloquium on Science, Arms Control, and National Security, "Science and Security: Technology and Arms Control for the 1990s," the Capital Hilton Hotel, Washington, DC, October 13–14, 1988. This event was sponsored by the Program on Science, Arms Control, and National Security of the American Association for the Advancement of Science.

The American Association for the Advancement of Science is grateful to The Carnegie Corporation of New York for supporting the publication of these articles and the colloquium for which this volume was produced. The statements and views expressed herein are those of the authors and do not necessarily reflect the views and policies of The Carnegie Corporation.

The AAAS Board of Directors, in accordance with Association policy, has approved publication of these articles as a contribution to the understanding of an important area. Any interpretations are those of the authors and do not purport to represent the views of the Board or the Council of the Association; its Program on Science, Arms Control, and National Security; or of the AAAS Committee on Science, Arms Control, and National Security.

AAAS Publication No. 88-21
Printed in the United States
ISBN 0-87168-336-9

Contents

Part I: The Defense Technology Base

Part II: U.S.–Soviet Relations

Acknowledgments

The production of this *Reader* was a collaborative process, and there are many to whom we are appreciative. Members of the Committee on Science, Arms Control, and National Security were very helpful in suggesting contributors to the volume. Iris M. Whiting, with the help of Sue O'Connell of the Publications Office, spent a *great* deal of time designing the format and copyediting this work. Susan Cherry designed the cover, and Raymond Orkwis arranged for its publication. Thanks also go to Julie Pek and Angela Hewitt for wordprocessing from the original copies, and to Matthew Budzik, Paul Carroll, and Douglas Russell for reading and rereading these essays. Tom Wander reviewed the final copy of the manuscript. We are also deeply appreciative of the efforts of the Soviet Embassy in obtaining articles from Soviet authors on U.S.–Soviet relations.

Finally, this book would not exist were it not for the high-quality, informative articles prepared by the contributors. Their responsiveness to our requests and ability to produce such expertly written pieces within a tight production schedule is indeed admirable. To them, the editor and those attending the colloquium are deeply indebted.

AAAS Committee on Science, Arms Control, and National Security

Contributors

Jim Blackwell is a senior analyst in the Advanced Systems Group at the Meridian Corporation. Previously, he served as co-chairman of the INF/START Verification Group from 1982 to 1987 and as a member of the U.S. delegation to the Mutual and Balanced Force Reduction Negotiations from 1974 to 1978.

Robert D. Blackwill is teaching foreign affairs, defense policy, and public management at the John F. Kennedy School of Government and is a senior research associate at Harvard University's Center for Science and International Affairs. Ambassador Blackwill has held several senior national security positions as a career diplomat in the White House and the Department of State, and was U.S. Ambassador to the Mutual and Balanced Force Reduction Negotiations in Vienna from 1985 to 1987.

Paul S. Brown is assistant to the Associate Director for Nuclear Design for Arms Control Matters at Lawrence Livermore National Laboratory.

Roger D. DeKok is the Director for Space Programs on the National Security Council staff. Col. DeKok has been a member of the U.S. Air Force on active duty since he received his commission as a distinguished graduate of the Reserve Officers Training Corps program at the University of Wisconsin in 1968.

Robert C. Duncan is the Director of the Office of Research and Engineering at the Department of Defense. Dr. Duncan is a former Vice President of Engineering for the Polaroid Corporation from 1969 to 1985, former Assistant Director of the NASA Electronic Research Center, and a former Director of Defense Advanced Research Projects Agency (DARPA).

Harold A. Feiveson is a research policy analyst at the Center for Energy and Enviromental Studies and an affiliate of the Center of International Studies at Princeton University. He has conducted extensive research on nuclear weapons and nuclear energy policy, and is co-principal investigator in a Princeton research program on nuclear policy alternatives.

Alexander Flax is the Home Secretary for the U.S. National Academy of Engineering. Dr. Flax is former President of the Institute for Defense Analyses, and former Assistant Secretary of the Air Force for Research and Development.

Jacques S. Gansler is currently Vice President of the Analytic Sciences Corporation and faculty member at the John F. Kennedy School of Government, Harvard University. Dr. Gansler was formerly Deputy Assistant Secretary of Defense and an industrial executive. He authored a book titled *The Defense Industry* (1980).

Alex Gliksman is conducting a study on emerging arms technology under a grant from The Carnegie Corporation of New York. His article draws upon this study's findings. He is currently writing a book for The 20th Century Fund on the implications of advanced conventional weaponry, and he previously conducted a two-year study on SDI and directed the U.S. Senate Foreign Relations Arms Control Subcommittee staff from 1982 to 1985.

Sidney N. Graybeal is Vice President of the Center for Strategic Policy at the System Planning Corporation. Prior to his current position, Mr. Graybeal was the Director of the Office of Strategic Research in the Central Intelligence Agency. He has also served in numerous U.S. government positions including: Chairman of the Arms Control and Disarmament Agency's (ACDA) SALT Backstopping Committee (1973–1976); U.S. Commissioner to the Standing Consultative Commission (1973–1976); Alternate Executive Officer for the SALT I negotiations (1969–1973); and Deputy Director of ACDA.

Jerry F. Hough is currently Professor of Political Science and Director of the Center on East–West Trade, Investment, and Communication at Duke University. Dr. Hough is also a senior fellow at the Brookings Institution and is the author of *Russia and the West: Gorbachev and the Politics of Reform* (1988).

Frank Jenkins is Assistant Vice President of the Science Applications International Corporation and Director of its Center for National Security Negotiations. While serving in the U.S. Air Force, Mr. Jenkins was assigned to the Joint Chiefs of Staff and conducted assessments of arms control proposals on strategic and theater nuclear forces, as well as serving as a military representative on several arms negotiation delegations.

Elizabeth J. Kirk is Senior Program Associate, AAAS Program on Science, Arms Control, and National Security. Dr. Kirk has been a member of the technical staff at the MITRE Corporation (1982–1988) and senior political scientist at Mathematica, Inc. (1975–1982).

Andrei A. Kokoshin is the Deputy Director of the U.S.A. and Canada Institute in Moscow and the Vice Chairman of the Committee of Soviet Scientists for Peace Against the Nuclear Threat. Dr. Kokoshin has written several books and articles on security issues in the Soviet Union, among them *Space Weapons: Security Dilemma* and *The Battle of Kursk in the Light of New Defensive Doctrines,* co-authored with General Larionov.

Michael Krepon is a Senior Associate at the Carnegie Endowment for International Peace, and Director of the Endowment's Verification Project. Dr. Krepon served in the Carter Administration, directing defense projects and policy reviews at the Arms Control and Disarmament Agency (ACDA). Mr. Krepon is the author of *Strategic Stalemate: Nuclear Weapons and Arms Control in American Politics*, and *Arms Control: Verification and Compliance.*

Michael MccGwire is a senior fellow in the Foreign Policy Studies program at the Brookings Institution. He is the author of *Military Objectives in Soviet Foreign Policy* (1987) and has another Brookings publication forthcoming, *Perestroika and Soviet National Security.*

Patricia Bliss McFate is President of the American–Scandinavian Foundation. Her former positions include Deputy Chairman, National Endowment for the

Humanities, as well as Vice Provost and professor in the School of Engineering and Applied Sciences, University of Pennsylvania. She is a fellow at the New York Academy of Sciences and was a member of the AAAS Committee on Science, Arms Control and National Security.

George H. Miller is the Associate Director for Nuclear Design at Lawrence Livermore National Laboratory. Dr. Miller also holds the position of Special Assistant to the Director for Defense Systems at the laboratory.

Vladimir A. Nazarenko is currently Deputy Editor in the Department on Disarmament Problems of the Novosti Press Agency, where he is an expert on politico-military and arms control issues. Col. Nazarenko has served on the General Staff of the Soviet Armed Forces, and is a Doctor of Military Science.

Milo Nordyke is a senior scientist in the Nuclear Test Program and was formerly the leader of the Treaty Verification Program at Lawrence Livermore National Laboratory.

William E. Odom is the Director of National Security Studies at The Hudson Institute. He recently retired as Director of the U. S. National Security Agency. Lt. Gen. Odom is the author of numerous books and articles on Soviet affairs, including *The Soviet Volunteers: Modernization and Bureaucracy in a Public Mass Organization* (1973).

The Office of Technology Assessment (OTA) was established in 1972 to provide nonpartisan analytical support to the U.S. Congress. OTA's basic function is to help legislative policy makers evaluate the consequences of technological changes and their applications.

Christopher E. Paine is a staff consultant to Senator Edward M. Kennedy (D-MA) and to Princeton University. He previously was a senior policy analyst for Physicians for Social Responsibility (1983–1985) and staff assistant for arms control for the Federation of American Scientists (1981–1983).

David B. Rivkin, Jr. is a Washington-based lawyer and defense analyst. He has written numerous articles, op-ed pieces, and books on various aspects of Soviet foreign and defense policies. His latest article was "The GOP Must Share the Blame for the Byrd Amendment," *Wall Street Journal*, June 15, 1988.

Sergei M. Rogov is a research fellow at the U.S.A. and Canada Institute in Moscow. Dr. Rogov has written many books and articles on international relations and American domestic and foreign policy and is the author of the forthcoming book, *A New Model for Soviet–American Relations*.

Josephine A. Stein was a consultant with the Office of Technology Assessment. She was formerly a Hewlett Fellow at Princeton University's Program on Nuclear Policy Alternatives and was a 1987 American Association for the Advancement of Science Fellow in Science, Arms Control, and National Security.

John D. Steinbruner is Director of the Brookings Institution's Foreign Policy Studies Program and a specialist on the Soviet Union and U.S. strategic policy.

Before coming to Brookings in 1978, Dr. Steinbruner held teaching and administrative posts at Yale and Harvard Universities and also taught at the Massachusetts Institute of Technology. He is the author of numerous articles and books in the field of Soviet–American relations and co-editor of *Alliance Security: NATO and the No-First-Use Question* (1983).

Karsten D. Voigt is ranking minority member, as well as speaker on arms control, for the German Social Democratic Party (SPD) in the West German Bundestag. He is also a member of the North Atlantic Assembly and the author of several books and articles on arms control and defense issues. Recently, Mr. Voigt testified at the U.S. Senate INF hearings.

Frank von Hippel is a Professor of Public Policy and International Affairs, Center for Energy and Enviromental Studies, Princeton University. Dr. von Hippel is also a former Chairman of the Federation of the American Scientists and is a member of the Board of Directors of the American Association for the Advancement of Science and the *Bulletin of the Atomic Scientists*.

Introduction

Elizabeth J. Kirk

As the presidency of Ronald Reagan draws to a close, we are struck by the changes that have taken place in the realms of national security and arms control over the past eight years. With an emphasis on maintaining the technical advantage, the Reagan era ends with the production of the B–1 and B–2 bombers, the Trident submarine, and the M–X missile, as well as the development of a scaled-down version of the Strategic Defense Initiative. The basing modes of the M–X or a possible smaller Midgetman, as silo-based or rail- or road-mobile, are being considered. Along with the new strategic weapons and many other high-tech systems come a plethora of cost overruns, test failures, and defense industry scandals. On the positive side the Packard Commission and others produced a series of recommendations for structural reforms necessary to make the defense systems acquisition process more effective. The high probability that the defense budget will not grow as fast as it did during the last eight years — and that it might even shrink — places demands upon the new administration to operate within a constrained budget. The advances brought about by high-tech weapons modernization must be tempered by national security policies which are realistically attainable to meet U.S. security needs.

These new demands are coupled with a dramatic shift in Soviet policies brought about by General Secretary Mikhail Gorbachev. His statements, believed by some to be empty rhetoric and by others to be a new friendly era for East–West relations, raise the possibility of radical shifts in Soviet military doctrine, openness towards data on Soviet military structures, a restructuring of the Soviet economy, and greater domestic freedoms for Soviet citizens.

This new environment is apparent in arms control, where, within the last year, the Intermediate-Range Nuclear Forces Treaty was signed and ratified, and new movement in the START negotiations took place in a series of high-level meet-ings. More generally, the Soviet Union has taken the initiative in presenting arms control alternatives instead of simply responding to alternatives proposed by the West.

The new U.S. president, therefore, will be faced with ongoing START negotia-tions and with the Conventional Stability Talks in Europe, intended to reduce con-ventional forces from the Atlantic to the Urals. He will also be faced with debates regarding the utility of a lower-threshold test ban agreement and Soviet on-site in-spectors in U.S. military installations. New verification methods, silo-based vs. mobile missiles, the appropriate applications of developing space weapons tech-nologies, and other security and arms control issues will be debated. In sum, the next four years provide numerous challenges and opportunities in the areas of na-tional security and arms control. Both East and West must seriously pursue these opportunities simultaneously for lasting changes to take place.

The issues in this volume are those that will be discussed at the Third Annual AAAS Colloquium on Science, Arms Control, and National Security titled, "Science and Security: Technology and Arms Control for the 1990s." These is-sues are organized in three major areas: the defense technology base, U.S.–Soviet relations, and arms control. The *Reader* and the colloquium are aimed at bridging the gap between the scientific/technical and political aspects of national security and arms control issues.

Part I contains six articles on the defense technology base of the United States. The Office of Technology Assessment chapter presents a cogent description of the defense technology base and some of the problems facing it today. The institutions involved in conducting defense research and development (R&D) and production are described. Robert Duncan's article outlines several basic issue areas that help drive and direct defense-technology-base considerations for the 1990s. Josephine Stein next examines the problems that have occurred with highly complex "ad-vanced" defense systems and how they may not, after all, contribute to our nation-al security. The bureaucracies and businesses involved in developing high-technology defense systems may have taken on a life of their own which may not always be directly tied to meeting security requirements. Alexander Flax dis-cusses the more specific relationships between defense R&D and the civilian marketplace. From his point of view, the problem lies not so much between defense R&D and its "payoff" to the U.S. economy as in the basic inability of the United States to turn technological advances into marketable products in a short period of time. Resolution implies modernization of U.S. production plants, a step many industries are unwilling or slow to take. Jacques Gansler examines in more detail the basic structural reforms, recommended by the Packard Commission and others, that are needed in order to streamline and improve the systems acquisition process. Finally in his essay, Alex Gliksman takes a look at the "revolution" taking place in high-tech conventional weapons and the proliferation of arms production capabilities.

Part II encompasses five articles on U.S.–Soviet relations. Sergei Rogov focuses on the establishment of positive U.S.–Soviet relations that go beyond the potential détente of the 1970s. He calls for a redirection of energies and resources toward cooperative foreign policy, military, economic, and social issues where both the United States and Soviet Union, as well as the rest of the world, would benefit. Jerry Hough contends that U.S.–Soviet tensions over the last forty years have had a stabilizing effect. He makes a plea, similar to Rogov's, for the United States and Soviet Union to focus on the opportunities for future cooperation, while not being afraid to make the sometimes radical changes necessary on both sides to bring them about.

The next three articles focus more specifically on U.S.–Soviet military relations. William Odom's article presents a historical description of the development of Soviet military doctrine and structures and the changes currently taking place as a result of technological advances and nuclear parity. Vladimir Nazarenko describes the current military strategies of the United States and the Soviet Union and calls for new strategies based on defense sufficiency and arms control. Finally, Michael MccGwire explains the reasons for the recent shift in Soviet military writing.

Part III contains several articles dealing with specific items on the arms control agenda. Frank Jenkins discusses the political and military requirements and ramifications of issues surrounding the strategic arms reduction talks (START). John Steinbruner and Andrei Kokoshin discuss the effects of deep cuts in U.S. and Soviet strategic forces. Robert Blackwill, Karsten Voigt, and David Rivkin each present their views on European security after the signing of the INF Treaty. Issues of maintaining or changing the nuclear deterrence strategy, conventional arms control, and the stability of the NATO and WTO alliance structures are also discussed.

Sidney Graybeal and Patricia McFate present a discussion of verification issues that have arisen in the arms control arena. Michael Krepon and Sidney Graybeal then outline in more detail the organizations and processes involved in current verification regimes. Finally, James Blackwell looks at areas where verification has proved difficult — mobile ICBMs, sea-launched cruise missiles (SLCMs), and air-launched cruise missiles (ALCMs).

Nuclear testing is next discussed. Feiveson *et al.* and Miller *et al.* cover the issues involved in monitoring nuclear tests at and below the current testing threshold limit of 150 kilotons provided for in the Threshold Test Ban Treaty. Here, the authors of the two articles have been asked to present arguments for or against new limits to testing. Feiveson *et al.* argue for a lower yield limit with monitored testing at a designated site. Miller *et al.* argue against lowering the limit since it diverts attention away from a more comprehensive approach to arms limitations and reductions. The final article, by Roger DeKok, outlines the Reagan space policy.

In all, these twenty-three articles provide an excellent outline of the major arms control, national security, and U.S.–Soviet foreign policy issues facing us today and well into the 1990s. They present the challenges and opportunities that science and technology present to us in the areas of national security and arms control.

Part I

The Defense Technology Base

1

What Is the Defense
Technology Base?*

Office of Technology Assessment

The defense technology base is that combination of people, institutions, information, and skills that provides the technology used to develop and manufacture weapons and other defense systems. It rests on a dynamic, interactive network of laboratory facilities, commercial and defense industries, subtier component suppliers, venture capitalists, science and engineering professionals, communications systems, universities, data resources, and design and manufacturing know-how. It includes laboratories run by the Department of Defense (DoD), other government departments and agencies, universities, and industrial concerns. It draws on the work of scientists and engineers in other nations. Information circulates both through formal routes dictated by chains of command, research contracts and other agreements, and through informal contacts within specialized technical communities, interdepartmental projects, and seminars.

Department of Defense technology base programs — and the accumulated results of these programs — are an important part of the defense technology base, but are far from all of it. Although DoD officials tend to speak of the defense technology base and the Department of Defense technology base programs interchangeably, they are not the same. The defense technology base is an accumulation of knowledge, skills, capabilities, and facilities, while the Defense Department's technology base programs are a collection of thousands of individual research

This article is the summary chapter (2) of a larger U.S. Congress Office of Technology Assessment report entitled The Defense Technology Base: Introduction and Overview — A Special Report, *OTA—ISC—374 (Washington, D.C.: U.S. Government Printing Office), March 1988.*

projects funded through the DoD budget, the results of which contribute to the defense technology base.

Almost all research and technology development can be drawn upon in producing defense systems, so that with the exception of classified research available only for defense applications, the defense technology base is largely the same as the national technology base as a whole. Of course, not all technology is of interest for defense applications, and not all is equally accessible to defense. Any research published in open sources (e.g., scientific journals) is available for use by engineers and scientists for defense applications. This includes foreign research and development, even work done in the Soviet Union. Proprietary work that is conducted by private companies, and remains unpublished in order to preserve competitive advantages, may also find its way into defense systems as those companies build the systems, subsystems, or components.

There are some practical limitations — amplified by recent government policy — on the transfer of technology between the defense and civilian sectors.[1] First, much defense technology is classified. Hence, it is only available to those working on defense projects. Second, researchers and engineers working on defense projects tend to form a community that interacts through mechanisms such as defense-related professional society meetings. Communication with those in similar fields doing nondefense work exists, but is often more limited. Indeed, in companies that do both defense and commercial work, engineers in either "side of the house" tend to be isolated from those in the other.

There are also mechanisms that reduce technology transfer and communication from nondefense areas to researchers and engineers doing defense work. Companies that develop commercial products seek to protect their investments by concealing their best technology as long as possible. Thus, cutting edge technology may remain inaccessible to DoD until after it has been introduced into the commercial marketplace. Additionally, some scientists and engineers prefer not to do defense-related work. Finally, regulations on doing business with the government tend to enforce a separation between companies that work for the government and those that do not — including separations between divisions of the same company. Knowing how to do business with the government creates a competitive advantage for some, while government regulations and contracting procedures present barriers against others. Indeed some observers argue that the problems of doing government business — close scrutiny, regulation of profits, and excessive military specification of product characteristics — tend to discourage innovative small- and medium-sized companies and steer them away from government work.

Thus, while in principle the defense sector can draw from a very wide technology base, there is some degree of isolation. Not all of that more general technology base flows into defense applications with equal ease.

DoD organizes its technology base programs into three categories which provide a working definition of the kinds of work and information that are considered part of the technology base. DoD's technology base programs consist of research into basic and applied sciences (funded under budget category 6.1), the exploratory development of practical applications of that research (budget category 6.2), and the building of prototypes to demonstrate the principle of an application (budget category 6.3A). Work funded under the remainder of the Defense Department's budget for research, development, test, and evaluation (most of DoD's RDT&E budget) is not part of the technology base.[2] In DoD jargon, "the tech base is 6.1, 6.2, and 6.3A," but the defense technology base is actually the accumulated results of those 6.1, 6.2, and 6.3A programs and much more.

Basic research, by definition, is almost entirely nonspecific in its potential applications. Most could lead just as easily to defense applications, commercial applications, or no practical applications whatsoever. There are, however, a few areas in which the Department of Defense has a specific interest in basic research because the connection to defense systems is clear. Examples are underwater acoustics (important for submarine detection and hiding) and the physics of explosive nuclear reactions (of obvious application to the nuclear weapons programs run by the Department of Energy [DOE]).

As science leads to technology, potential applications become clearer, and a sharper delineation of technologies with defense applications becomes possible. Some technologies are almost entirely military while others have little, if any, defense application. This separation is heightened as military programs become classified. Nevertheless, many technology areas are pursued for both military and commercial applications.

As technologies lead to the development of defense systems, developments that had been pursued primarily for commercial reasons can, and do, work their way into the defense systems. Prime contractors call on subcontractors for subsystems. Many of these lower-tier suppliers sell to both military and commercial buyers and use their commercial technology to develop products that are used in defense systems. Commercial components are sometimes designed directly into defense subsystems.

A diverse group of organizations contributes to the defense technology base. A large portion of basic research is performed at universities, which also train the next generation of scientists and engineers. University research is funded by the Department of Defense, other parts of the federal government (e.g., the National Science Foundation and the DOE, industry, and various private funds and endowments. DoD's university research program is growing and appears to have generated significant interest in the academic community. There appears to be more interest and capability than available funding permits in DoD support.

The U.S. Army, Navy, and Air Force maintain systems of research laboratories. Of the more than 140 individual laboratories, research and engineering centers, activities, and test facilities run by the Armed Services (Army, Navy, and Air Force), about half contribute significantly to the technology base. The rest concentrate on activities such as testing production aircraft. In addition to conducting research, some of these laboratories fund and monitor research by other organizations.

Other government laboratories — primarily the DOE national laboratories, the National Bureau of Standards, and NASA's research centers — contribute both directly and indirectly to the defense technology base. DoD contracts with the national laboratories and engages in cooperative research projects with NASA in areas of mutual interest. The results of research conducted at these institutions is generally available to organizations engaged in defense work.

A substantial part of the defense technology base is embedded in the defense industrial base and in the broader national industrial base. Much of the technology that finds its way into defense systems is developed by defense contractors and subcontractors and by commercial high-technology companies. In addition, some companies — e.g., AT&T, IBM, and UTC — run research laboratories that do a great deal of basic and applied research, most of which is available for defense applications.

The large defense contractors — e.g., Lockheed, Martin–Marietta, General Dynamics, McDonnell–Douglas, Rockwell — primarily design, develop, and produce weapons and other defense systems. They develop technology in-house and draw on technology developed elsewhere. They are both users of, and contributors to, the defense technology base. Because the defense industry sells only to the government, it operates under a special set of regulations.[3] And because there is only one customer (albeit one with many branches) and a limited type of competition, the defense market has evolved a unique set of business characteristics. There is controversy over whether this is the most efficient way to produce defense systems, and how to maintain sufficient capacity to meet surge requirements in the event of a conflict. These production issues are not a focus of this study, but the business structure of these companies strongly influences how they invest in technology. A more relevant issue is identifying the best methods for stimulating these companies to develop cutting edge technology for defense applications, draw on developments elsewhere, and incorporate the latest technology into products and production.

The industrial sector includes not just the defense industries that produce major defense systems, but also civilian industries. The so-called "dual-use" industries, which produce primarily for the civilian market, provide components for defense systems and stimulate technological advances that find their way into defense systems. Perhaps the best-known example of a dual-use industry is the semiconductor industry, the subject of a recent Defense Science Board study. In addition, laboratories run by companies that do very little defense work provide important basic technology that is eventually engineered into defense systems.

In some areas, civilian industries merely keep pace with or lag behind technologies that are being developed in the defense sector. But in other areas, it is the commercial firms that drive the pace of technological development. In general, the Department of Defense exerts strong influence on industries that are primarily devoted to defense and on newly emerging technologies. But DoD has far less influence with industries that have large commercial markets. For those industries, it is very much a minor customer; the civilian market shapes the industry and dictates the large investment in and consequent rapid progress of technical development. Defense production is a consumer of technology from these industries; defense interests are far less able to stimulate technology development in these industries than they are in the defense industries. If technologies that are dominated by the commercial market could not be transferred into defense applications, the defense sector would have to rely solely on technology developed in isolation, and would likely end up buying less-advanced technology than is available in the commercial marketplace.

A major focus of this OTA study is identification and evaluation of the factors behind the erosion of important dual-use U.S. industries at the leading edge of technology, and the implications of this erosion for national defense. The concern here is much less shaping technology development — defense is often a minority customer with only limited leverage on the industries — than it is ensuring that the technology and technical capacity will be available when needed. If these industries deteriorate substantially, the source of the technology will be in question, and if they leave the United States, DoD may find that the technology is no longer available and secure. Thus, DoD has a vital interest in the future of these industries. The government as a whole has an interest both from a national security perspective and, from a national economic perspective, although these two perspectives may not always coincide.

Issues

Maintaining this diverse defense technology base raises a large number of individual issues, which fall generally into the following seven categories:

- DoD's mechanisms for making technology policy and determining investment strategy to implement that policy;

- Funding for DoD technology base programs;

- The management of DoD laboratories and other government research institutions;

- Foreign dependence;

- Dual-use civilian high-technology industries;

- The defense industries; and

- The supply of scientists and engineers.

This section discusses these individual issues and the concerns from which they arise. These issues and concerns raise analytical questions that are not generally amenable to definitive answers, but provide a basis for analysis and informed debate. This section also presents — but does not analyze — some solutions that have been proposed. OTA reports these suggestions because they appear to merit exploration as Congress considers the issues, but OTA does not endorse them. OTA will explore some of these proposed solutions in further work on this project.

Department of Defense Mechanisms for Making Technology Policy and Technology Investment Strategy

The DoD science and technology program is a complex and sometimes bewildering array of 160 program elements encompassing thousands of individual projects, whose success is often difficult to judge. Consequently, there is widespread uneasiness that DoD may not be making the most effective use of its technology budget, and that its program may not be efficiently run. Some critics charge that technology base programs do not receive attention at a sufficiently high level, that Pentagon bureaucracies have no equivalent of a corporate vice president for research and development. Recognizing this problem, DoD has recently taken steps to address the situation. Other observers believe that the management system has developed the wrong focus: that performance is emphasized too highly over cost and quality, and product technology is emphasized to the virtual exclusion of process (manufacturing) technology.

There is also concern that "requirements pull" and "technology push" may be out of balance. Some argue that overly strict application of relevance tests in determining projects to be funded may be stifling creativity, while others point out that excessive loosening of the ties between research projects and military needs could lead to a technology base program that produces little practical benefit. There is an overriding concern that communications between developers of technology and military operators and planners are not sufficiently well developed. Developers could be more aware of military needs and planners could be more attuned to technological opportunities.

Research is by nature disorderly and risky. Its twin goals of seeking breakthroughs — including serendipitous, unanticipated discoveries — and evolving previous discoveries into useful applications are somewhat contradictory. Overconcentration on either is a prescription for disaster sooner or later. It may be that the apparent chaos of defense R&D programs is a reflection of these contradictions, and that these programs cannot and should not be managed in any more orderly fashion than they now are. Or it may be that valuable gains can be made through more effective management. Clearly, orderly evolution can benefit from orderly programs, but overly focused and controlled programs will inhibit the wide-ranging exploration that produces breakthroughs.

An area of growing concern is that of highly classified or "black" programs.[4] Congressional and bureaucratic oversight of these programs is very limited. Critics charge that black programs retard the diffusion and exploitation of important technology, while providing cover for poorly managed programs. Others claim that freedom from excessive oversight allows much more rapid progress and more efficient management. Some observers claim that technology transfers out of black programs slowly, if at all, but others claim that much of the technology in black programs came from "white" programs and only became highly classified when potential applications were identified.[5]

Some observers suggest that large-scale demonstration programs ought to be pursued as engines of technological innovation. They point out that projects such as Polaris and Apollo have served to focus the efforts of creative people and have produced much important technology. Like Apollo, such projects need not be confined to military goals. Some cite SDI and the National Aerospace Plane as current examples of such large-scale technology drivers. Others fear that such programs consume too much funding, driving less dramatic efforts out of existence, and that while providing a dramatic stimulus to technical and scientific development, they are an indirect and inefficient path to developing the technology that is desired. These critics argue that such programs take on lives of their own, and that as funding levels decrease, the secondary goal of spinning off technology is sacrificed to the primary goal of completing the program.

Each service runs its own R&D program, and the Office of the Secretary of Defense (OSD) coordinates these along with the efforts of the Defense agencies, the Strategic Defense Initiative Organization, and a few special projects. The Armed Services have different systems for setting R&D policy and for implementing that policy. These systems, and OSD's, have recently been reorganized. (Current organizations for making R&D policy are described briefly in a later section of this summary and in more detail in the main body of this report.) It is reasonable to ask whether each service's management system is optimizing its unique needs, or whether organizing them all along the same lines would be preferable, taking the best features of the three existing systems. Some observers believe that it would be useful to adopt management techniques used by other organizations that plan and manage R&D activities, such as private corporations and foreign governments. They claim that some of these organizations are better than DoD at setting and realizing technological goals and moving technology forward into products. Some in industry believe that DoD's systems suffer from the lack of a chief technical officer at a very high level and from an emphasis on managing programs rather than on setting policy.

Some observers argue that DoD's ability to attract and keep skilled management personnel is declining. They point out that skilled people can work effectively within a flawed system, but that even a perfect system cannot function well without top-quality people. They see these problems flowing at least in part from legislated restrictions on career paths, particularly those aimed at closing the "revolving door" between industry and government. People may be willing to sacrifice salary to serve their country, but are less willing to sacrifice their careers.

Congress faces some fundamental issues regarding technological base program management, including whether the DoD organization needs yet another shakeup or a radical new management approach. Even if the system is less than ideal, it may not make sense to reorganize it frequently instead of allowing it to settle down and do its job. Congress will be making implicit or explicit decisions regarding the extent to which it should be involved in detailed problems like organizing the DoD staff or selecting R&D programs and specifying their funding levels. Congress may wish to involve itself in the complex process of selecting technologies to be pursued and determining funding levels for each program. Alternatively, it may choose to limit its role to ensuring that the technology base programs have proper goals, adequate funding, and capable management.

DoD Technology Base Funding

Intimately tied to the issue of how DoD manages its technology base programs is that of the funding levels for those programs. This is a more immediate issue, since Congress wrestles with the budget each year. There are concerns that current levels may be inappropriate, and that within the overall totals, funding may be misallocated. Embedded in this latter concern is a worry that tech base program funding may not be adequately protected from "raiding" to support specific systems developments that are well beyond the tech base. Similarly, there is concern that large development programs like SDI or the Advanced Tactical Fighter tend to drain funds from technology base programs, and that the increasing emphasis on prototyping will be funded not with new dollars, but out of the existing technology base program. If the diversion of technology base budgets to support other programs turns out to be a significant problem, Congress may wish to consider taking measures to protect technology base program funding.

Gauging a proper level of technology base funding is difficult, as is the allocation of funding within that overall level. Since the output of a technology base program cannot be measured with any precision, the effect of adding or subtracting any particular sum of money cannot be calculated as it could be for programs such as procurement or maintenance. Some observers believe that it is most important to maintain a level of funding that is predictable and avoids dramatic fluctuations: constant changes in funding make it difficult to attract and keep staff.

The Strategic Defense Initiative (SDI) accounts for more than 40 percent of the technology base funding and almost all of the increase in technology base funding since 1981. But SDI is outside the system that controls the remainder of the technology base programs. Major changes in SDI could have implications for technology base funding as a whole, particularly since many programs that have more general utility are funded through SDI.

If Congress sets or endorses guidelines on basic R&D policy issues — such as the proper balance within the program between the orderly exploration and exploitation of known phenomena and the search for new breakthroughs in areas that are not yet recognized — is there a straightforward methodology that can be employed to determine the allocation of funding to implement those guidelines?

The Management of Government Laboratories

There is concern that the large array of government research institutions that contribute to the defense technology base does not form a coherent system to support technology needs. At issue is whether it could, and whether coherence is, on balance, desirable. There is also concern that some reorganization or consolidation may be in order, particularly among the DoD laboratories, and that they could function more productively if the organizations that operate them — or their relationships to those organizations — were changed.

Many of the Service laboratories are attached to systems commands with charters to develop specific classes of military hardware (e.g., airplanes, communications equipment, missiles). These laboratories are more like product centers run for their "customer" than they are technology centers. Others, like the Naval Research Laboratory, which the Navy views as a corporate lab and not a product lab, have greater latitude in their technical programs. There is concern that important areas of technology may be overlooked because of the narrow focus of parent organizations, and that innovative technologies that might ultimately benefit the overall mission of the command will be overlooked because they do not support the current major products of that command. Furthermore, unlike the contractor-operated DOE laboratories, DoD's laboratories appear to have a relatively difficult time shifting as the focus of technology shifts, or otherwise adapting to change. Others believe that a product focus is necessary and proper if the labs are to produce anything useful, pointing out that ultimately the task is to put systems into the field. They claim that many labs are not responsive enough.

The actions that Congress may wish to take will depend on understanding whether there are laboratories with unnecessarily redundant programs, substandard programs, unnecessary functions, or functions that could be performed more effectively elsewhere. Depending on the answers to these questions, Congress may wish to take measures to ensure more efficient use of government laboratories — e.g., closing, merging, or consolidating facilities; altering the command of Service laboratories; making greater use of non-DoD laboratories; e.g., setting up systems to enhance technology transfer.

Service R&D managers claim that it is becoming more difficult for the Service laboratories to attract and keep top technical talent, and that the United States may be risking deterioration of these important assets that have been built up over many years. Civil Service salary scales, never competitive with industry, are now in many instances not competitive with academia either. Aging physical plants are becoming increasingly less attractive relative to industry and academia. Finally, government service is becoming less prestigious. If these trends continue, are there measures that Congress could take to make defense laboratories more attractive to top scientists and engineers? Changing the management structure of these laboratories to allow compensation beyond that permitted under Civil Service rules has been suggested. This is being tried on a limited basis at the Naval Weapons Center (China Lake).

Foreign Dependence

This issue is intimately bound up with the next two — erosion of important civilian high-technology industries and problems in the defense industries. The United States is part of a global economy, particularly in high-technology industries. Foreign components are engineered into important defense systems, and other nations lead us in some areas of technology that are key to building defense systems.

This is, at least in part, a result of the success of post-war U.S. policy to build up the economies of friendly states. Many economists argue that the United States has no choice but to buy what it wants on a dynamic global market, and that to do otherwise will have major adverse effects on our own economy.

But we are caught on the horns of a dilemma. The least-risk approach from a national security perspective is to satisfy all our needs from domestic suppliers, so that our ability to get what we need is under U.S. control alone. This is, however, almost certainly not the least-cost approach. And it may ultimately risk losing access to important technology that either is not developed here or can only be developed here at a prohibitive cost or with a significant delay. Saving money through foreign sourcing, while generally appealing as a principle, can become much less attractive on a case-by-case basis when the specific economic impacts of funding a foreign project rather than its American alternative are examined.

Thus, some argue that the United States is becoming (or is in danger of becoming) too dependent on others for our defense technology. Others take the opposite position, that we are missing out by failing to take full advantage of the technological capabilities of our friends and allies.

Foreign dependence can be helpful and desirable, harmful and avoidable, or just unavoidable. In general, it is a mixture of all of these, complicating policy formulation. In a perfectly competitive world market, exploiting the technology of our allies would permit a more efficient division of labor, with each nation concentrating its technological efforts on what it does best rather than duplicating effort and spreading national resources too thin. It would allow us to exploit superior foreign technology where it exists by contracting for foreign technology development, licensing foreign technology, or buying foreign products. In the real world of today's marketplace, realizing these benefits may require taking measures to ensure that the U.S. share of this pie is maintained. Some observers have suggested that the United States make a serious effort to exploit foreign technology by establishing organizations to target, transfer, and exploit leading edge foreign technology. This might include reverse engineering of foreign products (i.e., analyzing them to understand how they are designed).

But this dependence risks cut-off supplies for either economic or political reasons. Foreign high-technology companies may be reluctant to commit resources to design specific items required by DoD when they find it more profitable to turn their energies in other directions. If the United States lacks a base of technical know-how in a particular technological discipline in which foreign sources cannot be induced to make developments in militarily significant directions, the Nation would be faced with a choice between stimulating domestic development or doing without the desired technology.

If other nations dominate industries based on technologies that are important to U.S. defense efforts, the United States will have to decide either to buy what it needs from foreign sources or to invest whatever is necessary to develop and keep a viable domestic technology and production base. This latter approach might involve inefficient measures to ensure domestic markets for domestic suppliers.

Ultimately, interdependence among nations may prove more advantageous to the United States than dependence on others. As the United States gets more deeply involved in reciprocal buy/sell relationships with its allies (particularly those with whom we have formal alliance ties), the risk inherent in relying on foreign suppliers is mitigated by mutual and interlocking military and economic dependence. Ties and interdependence are the political basis of alliance cohesion. This is quite qualitatively different from a situation in which the United States buys high-technology products in the international marketplace, and is at the mercy of the policies (or whims) of other nations.

Interdependence appears to be increasing in both the defense and commercial sectors. Within NATO, the United States participates in a number of groups working on cooperative development programs, and is involved in several major international development programs. The international nature of high-technology industry has led to interdependence among companies in various countries for components and finished products.

Congress is faced with the complex issues of whether the United States should (1) exploit foreign technology to a greater degree, (2) look for ways to reduce foreign dependence, or (3) encourage development of strategic U.S. high-technology capabilities, insulating them from foreign competition or at least preventing damage from unfair competition. A basic policy issue is how much the United States should become involved in cooperative defense programs, thereby increasing its dependence on allies and third parties for military technology. The United States will also have to decide how to respond to the movement offshore of high-technology industries that ultimately supply key technologies to defense systems and to foreign ownership of U.S. high-technology companies.

Resolving these issues will involve understanding whether, on balance, it is in the national interest to rely on friends and allies for selected military technologies and to allow those capabilities to diminish in the United States. This will, in turn, depend on how confident we are that our friends will remain our friends and technology will remain available. Fostering economic, political, and defense interdependence may help. It will also depend on developing criteria to identify those areas in which increased foreign dependence would be desirable or undesirable and on determining the risks to national security of certain technologies moving offshore. Ultimately, the United States will have to identify those technologies for which it is most important to maintain a domestic technology base. It may be that some technologies are so important to national security that the United States cannot accept foreign dependence under any circumstances. This may be particularly so in the case of technologies that are important to many systems. Some believe that dependence for any important military technology is unacceptable. Realistically, what are the options to stop this movement and/or to take compensatory measures that will preserve a domestic technology base despite market forces?

"Dual-Use" Civilian High-Tech Industries

The foreign dependence problem is most pronounced in those industries — e.g., semiconductors, computers, machine tools, structural materials, and optics — that are vital lower-tier suppliers for defense projects but do most of their business in the commercial marketplace. These industries are the sources of much of the innovation in their fields. But the Department of Defense has relatively little influence over them.

The degree of foreign dependence varies from industry to industry. In some, such as semiconductors and machine tools, foreign companies hold a majority of the market and control a major share of the technology. In industries like computers and materials, the United States still holds a decisive lead in the technology, but foreign companies are taking an increasing share of the market. As the market shares move to foreign companies, the technology is likely to follow. In the past, many major technologies began in the defense sector and expanded into commercial markets, reducing DoD's share and leverage. This pattern is still strong, although perhaps not as prevalent as it once was.

These industries are strongly influenced by the global market. If political or market forces move them toward foreign manufacture, foreign management, or foreign ownership, or away from developments of interest to defense, that is where they are likely to go. There is a two-fold danger in this. First, defense contractors may find it increasingly difficult to obtain the latest high-technology products. Some U.S. manufacturers believe that foreign suppliers of machine tools and computer parts sell U.S. companies products employing technology that is 2 to 3 years behind that of the products they sell their domestic customers. Second, DoD may find it increasingly difficult to induce those foreign companies that dominate a technology to use their R&D talents to develop specialized devices for military applications that are remote from most of their business. Ultimately, if the technical expertise moves entirely out of the United States, the services would have little idea what new technical developments they should request of the foreign suppliers. This latter stage would mean an obvious decline in U.S. technological leadership over the Soviets, and to some degree, a leveling of U.S. and Soviet access to the same advanced technology.

The government faces two problems with regard to these industries. First, from both defense and national economic perspectives, it faces the problem of keeping these industries viable here in the United States. Second, it has the problem of keeping the companies interested in doing business with the Department of Defense.

A number of rules and regulations pertaining to doing business with the Department of Defense inhibit companies from seeking government business and limit the flow of commercial products into defense systems. Innovative small- and medium-sized companies are particularly discouraged by conditions such as close scrutiny of their business, "red tape," and regulation of profits. Military specifications (MILSPECs) can have much the same effect, keeping out high-technology products that might be able to do the job but have not been designed to meet a complex list of specifications. These regulations by and large serve useful functions, but their utility should be weighed against the roles they play in keeping companies, and products, out of defense.[6] They have served to create a situation where experience in doing business with the government is an important company asset in competing for more government business, and lack of such experience is a major obstacle to companies trying to enter the market. Indeed, within some companies separations exist between divisions that sell to the government and those that do commercial business, stifling the transfer of technology within the company.

Most modern industrial nations have industrial policies and specific bureaucratic organizations responsible for industry. The movement of high-technology industries offshore can be attributed, at least in part, to other governments following policies that nurture these industries more effectively than do U.S. policies. Partnerships with industry and the availability of "patient capital," which encourages long-term results rather than short-term returns, are examples of the ways in which other governments encourage their industries. Developing and marketing high-technology products tends to be capital-intensive. The cost and availability of capital affect both the ability of companies to undertake long-term developments (which pay off handsomely but not soon) and their ability to rush a product to market once it is ready. The schedule for bringing a product to market can have a major effect on its share of the market. Some observers believe that the United States should be building the structure for a government-industry partnership to replace the current adversarial relationship. As Congress grapples with these problems, it will be important to understand the relationship between government policies and the deterioration of domestic high-technology industries that are important for defense. Are there realistic policy options that could reverse these trends?

This problem will raise before Congress the issue of whether the United States should have a more active national industrial strategy, and if so, what it should be, and what government agency should be responsible for it. It will be necessary to determine how defense needs would be taken into account and balanced against other needs in the formulation and implementation of that policy. Should the United States seek to maintain a viable domestic base for all industrial technologies that are of value for defense? Or would it make more sense to stimulate the international competitiveness of some U.S. industries and rely on the world market to meet the needs U.S. companies cannot provide? If the former is chosen, should the United States attempt to maintain, through government-funded projects, technical capabilities that are being lost in the commercial sector?

Defense Industries

For most technical developments of military significance, the road from laboratory to field runs through the large defense contractors. Each of these companies can perform a variety of important tasks, but their major roles are as prime system integrators, the companies that assemble the products of subcontractors and component developers into finished missiles, airplanes, and submarines. These companies, which both consume and develop new technology, have a unique niche in the economy, which results in unique business conditions.

There is long-standing concern that these conditions can inhibit technical development and efficient application of new technology. Like many corporations selling in the commercial market, defense contractors in the last few years have tended to plan for the short term rather than the long term. This de-emphasizes the value of investing in new technology which may only pay off 5, 10, or more years downstream.

Government procurement habits tend to reinforce this mode of planning. The method of government contracting has also tended to discourage both plant modernization and product innovation: the higher costs of operating inefficient plants can be charged to the customer, but the risk of failure due to trying new technology cannot. Much of the benefits from plant improvements — particularly large-scale improvements that pay back over relatively long terms — accrue to the government, further decreasing incentives for companies to fund such improvements. Of course, a company that can bid lower costs, because it will employ modern manufacturing technology or can produce a more capable product due to better technology, has an advantage in the competition for a contract. But it risks losses if it loses the competition despite investing in better technology. Congress may wish to consider whether government policies are inhibiting investment in modern manufacturing technology and what, if anything, should be done to stimulate these companies to develop and invest in better manufacturing technology. Related issues are whether (and how) government policies inhibit the transfer of technology to defense companies, discourage the entry of innovative companies into defense work, or encourage short-term rather than long-term planning. What options are available to correct these problems?

The IR&D (independent research and development) program has been controversial, especially within Congress, for many years. IR&D allows companies to recover some part of the cost of research programs initiated and funded by the company by treating it as a cost of doing business with, and developing products for, the Department of Defense. The cost is recovered as an allowable expense, similar to overhead, in contracts with DoD.[7] The companies decide what areas to work in and keep the commercial rights to their work. DoD judges the relevance of the work to defense needs and decides how much of the cost can be recovered, within an overall ceiling set by Congress. The remainder of the company's R&D costs are funded out of corporate profits.

In the view of the participating companies and other experts, IR&D benefits DoD because it allows the companies to stay current in areas of technology that are important to defense and it encourages the generation of research ideas by company scientists and engineers. IR&D is treated very seriously by most companies and receives high-level corporate scrutiny. Industry spokesmen point to long lists of specific systems that have originated in work funded through IR&D. The companies contrast this to contract research — which they also do — in which, for the most part, government officials decide what areas to pursue and exercise greater control over research projects. The companies are concerned that what they see as overcontrol by DoD will strangle IR&D. DoD, on the other hand, is concerned about keeping IR&D relevant to defense needs. The program tends to create friction between OSD — which supports the latitude for innovation that IR&D is supposed to provide — and the services, who want greater control over how their money is spent. Both DoD and industry agree that one benefit of the IR&D program is the communication it forces between government and industry researchers.

Congressional and other critics see IR&D as little more than a government giveaway, letting the companies bill the government for expenses they would incur anyway. They assert that the program is being abused. There is also concern that the incentives of the IR&D program are oriented toward short-term applications in which relevance can be demonstrated and cost recovered quickly, rather than to long-term advanced technology. One significant problem of IR&D is its complexity: even if all parties could agree that its goals are worthwhile, understanding the mechanisms of the program is very difficult. Some have suggested that the IR&D program be abolished in favor of simpler means to support company-initiated R&D.

Supply of Scientists and Engineers

Ultimately, a vibrant domestic defense technology base depends on a steady supply of highly capable scientists and engineers. Yet, in recent years, the rate at which U.S. citizens become scientists and engineers has fallen behind that of our allies and the Soviet Union. Furthermore, a large percentage of graduate students studying technical disciplines in the United States are foreign nationals, often supported by state and federal grants and subsidies. Although about two-thirds remain in the United States at least 5 years after completing their degrees, many go home once their education is complete (or a few years thereafter) and contribute their talents to foreign companies competing with U.S. high-technology companies. As long as these foreign nationals remain in the United States, they contribute to the technology base. The fact that so many choose to study in the United States is an indication of the quality of U.S. technological training.

This raises the issue of whether there is a need for additional scientific and engineering manpower — either across the board or in selected disciplines — to satisfy both national security and commercial needs, and if so, what steps Congress could take to increase the supply. These might include measures to increase the attractiveness of science and engineering careers to U.S. citizens and measures to increase the general quality of education. Congress may wish to consider the role of foreign nationals in the defense technology base and whether steps are necessary to influence the number of foreign graduate students in U.S. universities.

Exploiting the Defense Technology Base

An important element of getting new technology quickly into the field is how well the technology base can be exploited. Indeed, some observers believe that the major problems lie not in developing or maintaining the technology base, but in exploiting it, and that the major source of delay in getting technology from the laboratory to the field is not producing and developing the technology, but successfully designing it into military systems once it has been developed. Dr. Robert Costello, the Undersecretary of Defense for Acquisition, has cited the VHSIC program as an example of DoD's failure to get new technology rapidly into the field:

> While over 60 systems have VHSIC products in their plans, VHSIC parts are only in one current, deployed system... The technology is available and proven; and the Japanese are already using equivalent technology in commercial applications...[8]

Exploiting the technology base is not, strictly speaking, part of maintaining it. But in practice, the two are inseparable. No clear line separates developing technology from developing technology for applications. Part of creating a technology is developing the means to build something based on that technology. It is precisely in this area of manufacturing technology that many observers believe the United States has sustained the greatest losses in technological capability. They believe that the United States is not so much slipping in applied science and the development of product technology as it is in the craft and know-how of manufacturing devices: process technology must develop along with product technology if the product technology is to appear in actual devices. Moreover, designing a new technological advance into a useful device is itself a technology and may draw on several disciplines. For example, now that materials having superconducting properties at relatively high temperatures can be produced, the search is on both to find applications for these materials and to fabricate them into useful forms.

A related problem is that many program managers are reluctant to design new technology into their systems. Over the several years it takes a system to go through full-scale development and into production, the technology in some of its subsystems and components is likely to fall behind the state of the art, especially in areas that are evolving rapidly. Ignoring new technology results in a system that is behind the leading edge, but changing the system to include new technology will introduce delays. Adding new technology carries a risk that unforeseen factors will lead to additional delays. Thus, Service program managers have incentives to stay with the technology they know in order to minimize risks of schedule slippage.

One proposed solution is to insert new technology in retrofits to existing systems. Retrofitting takes less time than new developments and, therefore, gets the technology into the field faster. It also has the advantage of upgrading fielded capabilities without waiting for a new generation of major equipment.

A related approach is to include new technology in scheduled block upgrades of systems in production. Introducing new technology subsystems would not have to wait for the design and incorporation of an entire new generation of the major system. A company that is now building a particular fighter airplane might also be working on an upgraded design to which it would switch in a year or two. Both of these approaches are now used to some extent. Both require some degree of prior planning to ensure that new technology can be inserted with minimum disruption.

Yet another approach is through the organizational links between technology and systems development. For example, the Air Force had reorganized its technology base programs to be closely linked to the systems commands which are responsible for developing and buying equipment. This is an approach with attractive features, because it is supposed to make program managers more aware of new technology and make technology development more responsive to the needs of program development, thereby speeding the introduction of technology into systems. But linking technology too closely to development also risks stifling creativity — especially in areas that program managers do not currently recognize as being relevant to their missions.

How the Defense Department Makes and Implements Technology Policy

The Defense Department's ability to manage its technology base programs is central to maintaining the defense technology base. In fiscal year 1988, DoD invested $8.6 billion in technology base activities. (See Table 1.) One-third of the work was conducted "in-house;" more than half was contracted to industry, and universities performed the rest. Efficient management of the program — deciding what technology base policy is, allocating resources to implement that policy, avoiding unnecessary duplication of effort, and ensuring that important areas do not "fall between the cracks" that separate the elements of the program — will become all the more important if budgets become increasingly constrained.

Table 1. Department of Defense Fiscal Year 1988 Funding of Technology Base Programs (in millions of dollars)

	Army	Navy	Air Force	DARPA	Total
Research (6.1)	169	342	198	83	902*
Exploratory development (6.2)	556	408	557	512	2,033*
Advanced exploratory development	319	227	754	202	1,502*
Total services and DARPA					4,437*
Strategic Defense Initiative					3,604*
Other defense agencies					564*
Total DoD technology base programs					8,605*

*This sum includes $110 million for the URI which OSD has not yet allocated among the three Services and DARPA.

Source: Office of Technology Assessment, 1988, from data supplied by the Office of the Secretary of Defense.

Generally speaking, each service department assembles a technology base program to suit its particular needs. OSD exercises oversight and attempts to ensure that these programs are balanced and coordinated into a coherent whole with those of the defense agencies and the Strategic Defense Initiative Organization (which accounts for over 40 percent of DoD's technology base program funding). How all this works in detail is rooted in the structures of the various organizations within OSD and the srvices that make R&D policy, but is not entirely determined by them. These structures are still undergoing reorganization in the wake of the Goldwater–Nichols Reorganization Act (Public Law 99-433). Consequently, what follows may well change in the months ahead. Once the structures are settled, it may take even more time for the process to adjust to the structure.

The three services and the defense agencies (particularly the Defense Advanced Research Projects Agency [DARPA]) formulate their technology base programs with overall guidance from OSD. The Under Secretary of Defense for Acquisition has principal responsibility for all RDT&E activities except those of the Strategic Defense Initiative Organization (SDIO), which reports directly to the Secretary. For the services, the principal focus for this guidance and the principal point of contact is the Deputy Under Secretary of Defense for Research and Advanced Technology (DUSD[R&AT]), a deputy to the Under Secretary of Defense for Acquisition. After the Services have formulated their programs, the role of the DUSD(R&AT) is to structure the overall program across service lines in order to eliminate gaps and overlaps and enhance the return on investment. DUSD(R&AT) works continually with the services to help them achieve mutual interests and balance in their science and technology programs. DUSD(R&AT) coordinates activities with other government agencies and the scientific community.

Although DUSD(R&AT) is involved in coordinating service programs with those of DARPA and the other defense agencies, he does not have oversight for the technology base programs of the defense agencies. However, since DARPA contracts most of its programs through the services, much of its effort in fact receives DUSD(R&AT) oversight. The DUSD(R&AT) and the directors of all the defense agencies except DARPA report directly to the Director of Defense Research and Engineering (DDR&E), who reports to the Under Secretary for Acquisition. The Director of DARPA is also the Assistant Secretary of Defense for Research and Technology (ASD[R&T]) and reports directly to the Under Secretary for Acquisition.

DUSD(R&AT) runs some programs directly. The Computer and Electronics Technology Directorate of the DUSD(R&AT) manages the Very High Speed Integrated Circuit (VHSIC) program; the Microwave/Millimeter Wave Monolithic Integrated Circuit (MIMIC) program; and the Software Technology for Adaptable Reliable Systems (STARS) program.

The three service departments maintain structures for managing their technology base programs that are similar in some ways, but nonetheless have significant differences. Each has been planned to take into account the peculiar needs of the service, and each is rooted in its own history. Each of the services conducts an annual "top-down, bottom-up" planning exercise. From the top, each receives OSD's annual Defense Guidance Manual (as well as specific Service guidance), and from the bottom, each research institution contributes a technology base plan. Outside advisory groups and in-house technical directors and staffs also contribute. Planning includes a review and evaluation of the previous year's programs. It culminates with decisions to start new programs, continue or terminate existing programs, or transition programs into a new category (e.g., from 6.1 to 6.2).

Of the three services, the Navy has removed its technology base management institutions farthest from its procurement institutions. But relevance to Navy needs remains a powerful factor in selecting projects, particularly those beyond 6.1. Although it has the smallest overall technology base program, the Navy has the largest research (6.1)program and performs the largest fraction of the work in-house. The 6.1 and 6.2 programs are run, respectively, by the Office of Naval Research and the Office of Naval Technology, both of which report to the Chief of Naval Research, who in turn reports directly to the Chief of Naval Operations (CNO) and the Assistant Secretary of the Navy for Research, Engineering, and Systems. Some Navy laboratories are run by the Office of Naval Research and others are run by the Space and Naval Warfare Systems Command (SPAWAR), which also reports directly to the CNO. Until recently, the laboratories now run by SPAWAR were run by the "buying commands" such as the Naval Air Systems Command and the Naval Sea Systems Command. The decisions regarding 6.2 work to be done in SPAWAR laboratories are made by the Office of Naval Technology. Of the three Services, the Navy has the smallest advanced technology demonstration (6.3A) program; it is run directly by the Office of the CNO (OPNAV).

In contrast to the Navy's program, which is structured for finding and developing new technology of use to naval missions, the Air Force structure puts greater emphasis on getting technology into the field. The Air Force 6.3A program, which the Air Force sees as transition money, is five times as large as the Navy 6.3A budget. The technology base programs are run by Systems Command, the individual divisions of which are the Air Force's "buying commands." These divisions run the laboratories which conduct and manage the 6.2 and 6.3A programs. Roughly two-thirds of the work is contracted out. The 6.1 programs are administered by the Air Force Office of Scientific Research, which is also part of Systems Command. Within Systems Command, the Deputy Chief of Staff for Technology and Plans has oversight responsibility for the science and technology programs, but these programs must be approved by the Director of Science and Technology Programs in the Office of the Assistant Secretary of the Air Force for Acquisition. The Air Force has recently taken a decision to treat its technology base program as a "corporate investment" that deserves a fixed fraction of the total Air Force budget.

The Army's structure is perhaps the most complicated and decentralized of the three services. The Army has the smallest headquarters staff and relies the most on elements outside headquarters to run the technology base programs. Some aspects of the Army system are analogous to the Air Force system, while others are similar to the Navy's. About three-fourths of the science and technology programs are run by the Army Material Command (AMC); other S&T programs are run by the Surgeon General of the Army, the Corps of Engineers, and the Office of the Deputy Chief of Staff for Personnel. The offices within these organizations that are responsible for the S&T programs all report for oversight by Army headquarters to the Deputy for Technology and Assessment who, in turn, reports to the combined Office of the Assistant Secretary of the Army for Research, Development, and Acquisition (RD&A) and the Deputy Chief of Staff for RD&A. Within the Office of the Deputy for Technology and Assessment, it is the Director of Research and Technology who has responsibility for planning and coordinating the entire technology base program. The Army Materiel Command is analogous to the Air Force Systems Command.[9] It is organized into mission-specific "buying commands," which run laboratories and technology base programs, and Laboratory Command (LABCOM) which, like the Office of Naval Research, runs laboratories and conducts generic research. Under LABCOM, the Army Research Office is responsible for AMC's 6.1 programs. Unlike the Office of Naval Research, the Army Research Office has no in-house capabilities and contracts all of its work out.

Roles of the Component Research Institutions

Nearly 200 federal government laboratories, research and engineering centers, activities, and test facilities contribute to the defense technology base. About 65 of these play a major role. Of these major centers, roughly 50 belong to the Army, Navy, and Air Force Departments, and the rest are run by the Department of Energy, the Department of Commerce, and the NASA. Many of the DOE laboratories are government owned contractor operated (GOCO) facilities, permitting the hiring of personnel outside the Civil Service system. The defense agencies do not operate laboratories; however, DARPA maintains a close liaison with some service laboratories.

Army

Within the Army Materiel Command, six "mission-specific" systems commands run research, development, and engineering centers that conduct exploratory and advanced technology development into areas that are specific to the command mission, while the Laboratory Command(LABCOM)operates laboratories that are focused on generic research and development, primarily at the 6.2 and 6.3 levels. The Army's research (6.1) program is almost entirely contracted out. Some of the mission-specific centers are more influential than LABCOM and greatly influence the structure and priorities of the Army S&T program.

LABCOM runs the Electronic Technology and Devices Laboratory, the Materials Technology Laboratory, the Human Engineering Laboratory (the Army's lead lab for man-machine interface and robotics), the Ballistic Research Laboratory, the Atmospheric Science Laboratory, and the Vulnerability Assessments Laboratory. It also runs Harry Diamond Laboratories, which is involved in a variety of areas including ordnance electronics, electromagnetic effects, and advanced electronics devices. These laboratories do most of their work in-house, but contract out a substantial portion.

The facilities belonging to the systems commands conduct some exploratory development, but the main thrust is toward the "mature" end of the S&T program, leading to components, products, and systems. These centers conduct some in-house work, but the bulk is contracted out. For the most part, they bear the names of their parent commands and their missions are specified by their names: the Armaments, Munitions, and Chemical Command; the Aviation Systems Command; the Missile Command; the Tank Automotive Command; the Communications Electronics Command; and the Troop Support Command.[10] The Missile Command Research and Development and Engineering Center has the capability to carry a concept nearly to production almost without outside help. The Night Vision and Electro-optical Laboratory, run by the Communications Electronics Command, is a recognized leader in infrared and other night vision devices for all three Services.

Navy

The in-house R&D capabilities of many Navy laboratories are substantial, through the tech base and into full-scale engineering development. Like the Army's Missile Command Research and Development and Engineering Center, some can carry a system through development and almost to production. At least one Navy laboratory — the Naval Research Laboratory — does a substantial part of its work as the result of proposals by lab personnel to conduct R&D for the other Services and other parts of the government. The Naval Weapons Center has developed major missile systems to the production stage.

The Office of Naval Research oversees the activities of several laboratories — including the Naval Research Laboratory, the Naval Ocean Research and Development Center, and the Naval Environmental Prediction Research Facility. The Naval Research Laboratory is the Navy's principal research laboratory and, in some areas, DoD's. It conducts a broad-based program that includes such diverse fields as computer science, artificial intelligence, device technology, electronic warfare, radar, materials, directed energy, sensor technology, space technology, and undersea technology.

Aside from ONR's laboratories, all Navy development centers and activities are assigned to the Space and Naval Warfare Command. Although these facilities perform 6.2 and 6.3A work, the emphasis is on development. The major centers are the Naval Air Development Center, the Naval Ocean Systems Center, the Naval Weapons Center, the Naval Ship Research and Development Center, the Naval Surface Warfare Center, the Naval Undersea Systems Center, and the Naval Coastal Systems Center. Science and technology programs in these centers include electro-optics, acoustics, microwaves, artificial intelligence and knowledge-based systems, ocean science, bioscience, electronic materials, structural materials, ship magnetics, hydrodynamics, and charged particle beams.

Air Force

All of the Air Force laboratories are run by the systems divisions of Systems Command — Armaments Division, Aeronautical Systems Division, Electronic Systems Division, Space Division, and Human Systems Division — in support of the division's specific mission. Exploratory and advanced development (6.2 and 6.3A) are funded directly by the divisions, while research (6.1) is funded and managed by the Air Force Office of Scientific Research. The centers are the Air Force Armaments Laboratory, the Rome Air Development Center, the Geophysics Laboratory, the Air Force Weapons Laboratory, the Air Force Astronautics Laboratory, the Human Resources Laboratory, the Aeromedical Laboratory, and the Wright Aeronautical Laboratories. The Wright Aeronautical Laboratories are a cluster which includes the Aeropropulsion Laboratory, the Avionics Laboratory, the Flight Dynamics Laboratory, and the Materials Laboratory. Managed by outside Systems Command, but coordinated with it, is the Air Force Engineering and Services Laboratory.

Department of Energy

The Department of Energy's laboratory system, established during World War II to develop nuclear weapons, has grown dramatically in scope and size. About 60 institutions employing 135,000 people and having a replacement cost of about $50 billion conduct research into physics, chemistry, cosmology, biology, the ecosystem, geology, mathematics, computing, and medicine, as well as a broad range of technologies that spring from these disciplines. Nine multiprogram laboratories and 30 specialized laboratories are involved in these fundamental science and technology activities. These laboratories maintain close ties with academic researchers, often subsidizing part of their work at the DOE facilities.

A primary focus of DOE's work for national defense is the nuclear weapons programs. Moreover, some of the major multiprogram national laboratories — Sandia Laboratory, Los Alamos Laboratory, Lawrence Livermore Laboratory, and Oak Ridge Laboratory — do 10 to 15 percent of their work under contract to DoD on defense-related research not directly related to nuclear weapons, and most of their work is available for exploitation by DoD or its contractors. Several other laboratories — Argonne, Brookhaven, Lawrence Berkeley, Idaho National Engineering, and Pacific Northwest — do little or no work for DoD, although they do basic and applied research that is available for exploitation for defense purposes.

The National Aeronautics and Space Administration (NASA)

NASA runs a number of major facilities, many of which are dedicated almost exclusively to the design and testing of space-launch and space systems. Several others, while they also support the space program, have a broader mandate. The major NASA field centers are the Ames Research Center, the Lewis Research Center, the Langley Research Center, the Goddard Space Flight Center, the

Marshall Space Flight Center, the Johnson Space Center, the Kennedy Space Center, and the Jet Propulsion Laboratory. These centers conduct work in aeronautics, space, communications, propulsion, and computers, among other fields.

Other Federal S&T Activities

By far, the majority of federal science and technology activities with potential defense applications are conducted within DoD, DOE, and NASA. There are pockets of activity within other agencies that can, and do, contribute to the defense technology base.

The National Science Foundation (NSF) funds three times as much university research as DoD. But NSF conducts no research of its own and does not contract for research; it funds unsolicited proposals. NSF supports research projects in almost every area of science and technology.

The Commerce Department's National Bureau of Standards (NBS) conducts a number of applied research projects that contribute to the defense technology base. NBS researchers, in partnership with others from industry, government, or academia, investigate areas such as electronic technology, information processing, biotechnology, chemistry, and manufacturing technology. This partnership arrangement is designed to get the technologies quickly into applications. NBS technical work is carried out in the National Measurement Laboratory, the National Engineering Laboratory, the Institute for Computer Sciences and Technology, and the Institute for Materials Science and Engineering.

The Department of Transportation — particularly the Federal Aviation Administration and the Coast Guard — the Department of Agriculture, the Department of Interior's U.S. Geological Survey, and the National Oceanic and Atmospheric Administration all share interests with DoD and conduct some research and development with potential defense applications.

University Research Programs

University laboratories perform a substantial portion of the basic and applied research conducted in the United States. They work in all areas of basic research and most areas of applied research and exploratory technology. University research is an area of potentially high leverage for DoD, since DoD funds less than 25 percent of it, but has access to almost all of it. DoD has generated great interest in its new programs for university research; many more requests have been received than could be funded. The two most significant are the University Research Instrumentation Program, to upgrade laboratory equipment, and the University Research Initiative, most of which goes to multidisciplinary research programs and programs to promote scientific training. DoD has also put money into creating university "centers of excellence" in selected disciplines.

NOTES

1. For example: Department of Defense Directive 5230.25, Nov. 6, 1984; and Executive Order 12356. For more detail, see Science Policy Study Background Report No. 8, *Science Support by the Department of Defense*, prepared by the Congressional Research Service for the Task Force on Science Policy of the House Committee on Science and Technology, December 1986.

2. Strictly speaking, by the Office of the Secretary of Defense definition, the technology base programs are 6.1 and 6.2. Technology base plus 6.3A are the science and technology programs. However, these definitions are often used interchangeably, and in recent years, common usage has been to refer to 6.1, 6.2, and 6.3A as technology base programs, while 6.3B and 6.4 are specific system developments linked closely to procurement.

3. Those divisions of defense companies that sell in the civilian marketplace operate differently.

4. See, for example, Alice C. Maroni, "Special Access Program and the Defense Budget: Understanding the Black Budget," Congressional Research Service, Issue Brief IB87201, Dec. 2, 1987.

5. Of course, these contentions can be only verified by those with access to the black programs. However, both are logical. If few people know about a project, there will be few opportunities to envision other applications for the technology. Very little technology is born highly classified; it only becomes worthwhile to limit access to information when its potential applications have been identified.

6. Some in industry see many of these regulations as unfair.

7. The results of IR&D recovery (reimbursement) improve a company's competitive position by improving its technology base, but the addition to the company's cost basis reduces its competitiveness on future contracts.

8. Quoted in *Defense Daily*, Dec. 3, 1987, p.203.

9. The Navy equivalent, the Naval Material Command, was eliminated a few years ago in a reorganization. The component commands — e.g., the Naval Air Systems Command — now report directly to the CNO.

10. For example, the Armaments, Munitions, and Chemical Command runs the Armament Research, Development, and Engineering Center and the Chemical Research, Development, and Engineering Center.

2

The U.S. Defense Technology Base: Issues for the 1990s

Robert C. Duncan

The importance of technology in defense was demonstrated as long ago as 212 B.C., when it is said that Archimedes used mirrors to concentrate the sun's rays on the Roman fleet besieging Syracuse, setting Roman ships on fire. It took some 2,000 years more to advance from collimated, incoherent light to highly focussed laser light, as reported by Arthur Schawlow and Charles Townes in their landmark paper "Infrared and Optical Masers," published in the December 15, 1958 *Physical Review*. This research had been supported under the Joint Services Electronics Program at Columbia University.

As the pace of technology quickens, new weapons systems provide vastly increased capabilities, but have grown more complex, taking longer to develop and greater skill to maintain, thus increasing both developmental and operating costs. Norman Augustine, in his book *Augustine's Laws*, plots the cost of aircraft as a function of time and comments (tongue in cheek) that some time early in the twenty-first century, the country will be able to afford only one airplane. To avoid dilemmas such as this raises many issues. They are the subject of this paper.

Selecting Technology Base Programs

For the next few years, technology base programs will be selected under severe budget constraints, making the system less forgiving of poor choices. Long-term goal setting must be followed by strategic choices. Since this year's manager may not commit next year's incumbent, consensus building becomes a central issue if future managers are to continue to feel a commitment to even the best-laid plans. Further, goals and plans and choices effectively communicated to a well-trained work force will improve motivation, quality, and productivity.

The technology base, as used here[*], consists of three categories of research and devlopment (R&D), known by the first two digits of their budget category numbers: 6.1 for (basic) research, 6.2 for exploratory development, and 6.3A for advanced technology development ("A" because it is only a fraction of the entire 6.3 line, advanced development). The Strategic Defense Initiative (SDI) is all in 6.3A, but since it is in a class by itself, the figures that follow will not include SDI unless otherwise stated. For fiscal year 1988, the budget numbers in billions of dollars were: for 6.1, $0.9 billion; for 6.2, $2.4 billion; for 6.3A $1.9 billion ($5.4 billion with SDI). The total technology base funding for fiscal year 1988 is $5.2 billion ($8.7 billion with SDI).

Central to planning and execution of a viable technology base program is the definition of programs and projects to be undertaken. It is prudent and necessary that the Department of Defense (DoD) conduct a broad technology base program in areas of special needs that are not otherwise available from industry, academia, and non–DoD government laboratories. Much of this activity is evolutionary and plays an essential role in the formulation of our future technological posture. In addition, there is a need to identify and emphasize technologies that provide the high leverage needed to enhance deterrence.

Nuclear Verification and Monitoring — Third World Threat

The paramount threat to our national existence has been, and continues to be, the ability of the Soviet Union to deliver the enormous destructive power of nuclear weapons. This threat has been somewhat reduced, but not eliminated, by the recent agreement with the Soviets on Intermediate-Range Nuclear Forces (INF), which was, in large part, due to the R&D carried out under SDI. These developments have increased the need for effective strategic verification measures.

In the meantime, other nations are seeking or acquiring a nuclear weapons capability. Some of these countries possess, or may soon possess, intercontinental delivery systems to go with their nuclear weapons. This threat from Third World powers complicates the nuclear equation and requires continual assessment of the nuclear threat. Ultimately, this may call for the development of new strategies and technologies to deal with a nuclear-proliferated world.

---

[*]   In budget terms, "Technology Base" comprises only 6.1 and 6.2. The sum total of 6.1, 6.2, and 6.3A is then referred to as "Science and Technology."

Conventional and Nonnuclear Battles

The immediate Soviet threat has thus shifted (at least temporarily) away from the theater nuclear arena toward conventional capabilities. This is reflected in recent DoD initiatives such as the "Conventional Defense Initiative" and the "Balanced Technology Initiative" of 1987, which are analogous to the "Strategic Defense Initiative" of 1983–1984 that responded to the strategic nuclear threat. We must not forget that the maintenance of a robust conventional deterrent is manpower intensive and very costly. In many respects, the nuclear deterrence that has prevented global warfare for so long has been deterrence on the cheap. The purchase price has consistently been a relatively modest share of all the resources allocated to defense. Therefore, in addition to developing improved conventional weapon systems, technologies that reduce demands on service personnel will be at a premium.

The newer conventional defense initiatives address the needs of tactical warfare, which is virtually endemic in many parts of the globe. In recent years, hostilities have flared in Latin America, the Middle East, Africa, and Southeast Asia. Equipment made in the United States has been used in diverse regional conflicts in Afghanistan, the Iraq–Iran war, and the Falkland Islands. Sometimes, our weapons have been pitted against Soviet-made weapons, and we have learned from the experience. Future wars involving U.S. or allied troops may be fought without a conventional battle front; they may be a series of actions in depth, more akin to large-scale unconventional warfare, where good communications and high mobility confer a tremendous advantage on the forces possessing them, especially when simultaneously denied to the enemy. Noncombatants may also be present on the battlefield or its vicinity, hence the need for "foolproof" methods of noncooperative classification and positive identification of potential targets.

Deputy Secretary of Defense William Taft recently reminded us that "the need for conventional defense improvement has reached a level of unprecedented urgency." By the application of technological advances, our forward-echelon capabilities have been improved through the Army's modernization program and the introduction of systems such as the M–1 tank, the Multiple Launch Rocket System, and the Apache helicopter. Tactical capability in the air has been strengthened by the introduction of the F–15, the F–16, and the F–18, accompanied by advances in "smart" air-to-air and air-to-ground munitions. The so-called Assault Breaker systems should do much to help us locate and destroy the enemy's forces well behind the forward edge of the battle area (FEBA), thus disrupting an aggressor's attack plan.

For the forward-echelon forces, technology advances will improve armor/anti-armor capabilities; the Defense Advanced Research Projects Agency (DARPA), the Army, and the Marines have undertaken a new initiative to remedy deficiencies in this area identified in a recent Defense Science Board study. For support of forward-echelon forces, other promising technologies include electromagnetic guns, mine/countermine warfare, and brigade-level reconnaissance, surveillance, and automatic target acquisition and recognition with elevated sensors. These capabilities are due, in large part, to advances in technology, which must be continued if we are to overcome an inferiority in numbers.

Terrorism — Detection and Prevention

The rise in terrorism gives us pause when we try to comprehend the consequences of large means of destruction falling into unscrupulous hands. Terrorism poses both physical and psychological hazards that are difficult to combat. Detection and prevention is the best defense, since the object of the terrorist is usually blackmail and publicity. Hence, there is a need for better arms control technologies, from better inventory keeping to improved surveillance, detection, and identification.

Scientific and Technical Manpower

The need for skilled manpower will grow as weapons and support systems incorporate more technology, just when the baby boom is over. The Army Research Institute for the Behavioral Sciences has estimated that by the year 1995, to maintain the current force structure, the Army would have to recruit 75 percent of the eligible (17–21 years old) manpower pool with a propensity to join the Army — unless something is done about it. (This assumes that women will not replace men and that surveys taken to determine "propensity," that is, willingness to seriously consider joining the Army, remain valid.) Technology may have to come to the rescue. It can devise better training and trainers, such as, more realistic simulators; provide computer aids, such as knowledge-based systems to aid in diagnostics; and provide smarter built-in test equipment to predict failures before they occur.

Technology Opportunities

The impact of new technologies may be revolutionary or, just as important, evolutionary. Technological progress may improve product quality, enhance existing capabilities, or (as has happened with nuclear weapons), revolutionize warfare. Each is important in its own way, and each has its special role to play. High-temperature superconductors and biotechnology could result in revolutionary approaches to old problems. Automation and knowledge-based systems will create many opportunities, plus savings in materiél and manpower. More fuel-efficient engines and lighter-weight materials will affect warfare through higher mobility of forces. Software productivity, reliability, and convertibility are serious limiting factors today, waiting to be resolved. Manufacturing, logistics, and construction — and all the support technologies without which an army cannot function — are ripe and ready for the applications of new technologies with potentially great impact, perhaps waiting for new disciplinary approaches to educate future specialists in those fields. The environmental and life sciences also have great impact in war as well as in peace. We cannot afford to overlook any technology opportunities. I am confident that we have the intellectual resources to tackle these challenges and opportunities.

Planning a Defense Technology Strategy

Under Secretary of Defense for Acquisitions Robert Costello, in his testimony to Congress on November 19, 1987, submitted a ten-point program of his goals and strategies, which may be summarized as follows:

- Bolstering industrial competitiveness;
- Initiating total quality management;
- Developing a strategy for international technology;
- Encouraging competition;
- Reforming acquisition regulations;
- Integrating acquisition information systems;
- Reducing technology lead time by 50 percent;
- Establishing mutual trust between industry and DoD;
- Increasing the effectiveness of the technical work force; and,
- Increasing involvement of small business.

The technology base issues discussed in this paper relate to many of these goals.

Quality — Total Quality Management and Competitiveness

Weapons systems (or other products) capable of the highest performance are useful only when they satisfy criteria of *affordability, dependability,* and *timeliness.* Any technology strategy must explicitly address these "ilities." Accordingly, Costello formulated a program of Total Quality Management. This requires close cooperation with practitioners and researchers from industry and academia, supported by dedicated technology programs to develop new processes for strategic design, manufacturing, logistics, and information management.

The Air Force's "Forecast II" study in 1986 listed future high-payoff technologies, among them "Unified Life Cycle Engineering" to integrate the entire process from design to manufacture and beyond ("strategic design") by paying attention to issues such as dependability, maintainability, and operability as early as possible in the life cycle of the product. A similar approach being investigated by DARPA in 1988 is called "Concurrent Design"; in addition to integrated planning and modelling, it would also rely on early experimentation to test critical design features.

The Navy is building a flexible manufacturing facility, and supports the Advanced Manufacturing Research Facility at the National Bureau of Standards. The National Science Foundation has treated design theory as an engineering discipline and recommended a "strategic manufacturing" initiative. DoD is coordinating its activities with concerned government agencies, industry, and universities, for example, strongly supporting SEMATECH (the Semiconductor Manufacturing

Technology Center) at Austin, Texas [recommended by the Defense Science Board (DSB) in its February 1987 report], and the National Manufacturing Science Center at Ann Arbor, Michigan.

Technology Transition — Reducing Technology Lead Time

Technology insertion was a principal goal of then-Deputy Secretary (now Secretary) of Defense Frank Carlucci's P^3I (pronounced P–cubed–I, for "pre-planned product improvement") 1981 management initiative. His purpose was to avoid delay in product development by using known technology, while allowing for the insertion of still developing technology later on.

A Quest for Excellence, the 1986 report of the President's Blue Ribbon Commission on Defense Management (the Packard Commission), addressed the same theme, strongly recommending the use of prototypes in the acquisitions process to determine to what extent a given new technology can improve military capability and to provide a basis for making realistic cost estimates prior to a decision on full-scale development. The commission further recommended that DARPA should stimulate greater emphasis on prototyping by conducting prototype projects that embody technologies that might be incorporated into joint programs or into selected service programs. DARPA is currently conducting two prototype programs, one Army and one Navy, which include the preparation of plans for subsequent full-scale development.

The DSB undertook a study in the summer of 1987 on technology base management and came up with three recommendations, one of which concerned technology transition. The DSB proposed a program of "advanced technology transition demonstrations" (ATTDs) to be initiated within the Advanced Development program. The principal objective of the ATTDs is to build and test experimental systems in a field environment before making full-scale engineering development decisions with user participation.

The DSB study recommended the application of selection criteria and management principles that have proven effective during past technology development and demonstration efforts; that by 1991, at least half of Advanced Technology Development (6.3A) funds be directed to ATTD projects; and that ATTD projects be reviewed annually by the Vice Chairman of the Joint Chiefs of Staff to ensure that projects address user needs.

In 1987, the Air Force started a technology insertion program to increase reliability, maintainability, and survivability. It is called RAMTIP (Reliability and Maintainability Technology Insertion Program). This full-scale engineering development program uses more mature technologies than would be the case under ATTDs. The ATTD concept would attempt to match technologies at an earlier stage of development with systems that could benefit from incorporating them.

Cooperation and Synergism Between Military and Civilian Sectors, and the United States and Its Allies

Their research centers and other initiatives bring together civilian and defense related R&D activities where their goals coincide. Indeed, military and economic activities are interdependent, as pointed out by Paul Kennedy in his widely discussed book, *The Rise and Fall of the Great Powers*. Since defense must of necessity be at the cutting edge of technology, technologies supported by the military have always contributed to scientific and technical progress in general, with many spinoffs to civilian applications. Thus, the defense technology base has advanced our industrial competitiveness by developing technologies from aeronautics to X-ray lithography and has made nearly all of them available to industry. Conversely, defense needs a strong industrial base.

Technology opportunities produced by both civilian and military R&D abound and only await exploitation. Advances in microelectronics could soon enable us to to put the computing power of a mainframe computer into a missile guidance system. Advances in sensors and machine intelligence could also enable us to automate the search for and recognition of targets in a scene, items in a warehouse, or, for that matter, tools on a factory floor.

In 1986, Congress passed the 1986 Federal Technology Transfer Act, the purpose of which is to enable the federal R&D establishment to play a more active role in technology transfer. The Act provided for cooperative research activities between federal laboratories and private companies. It reinforced the notion that military and economic progress are intertwined.

To make the most of our resources, we work with our allies wherever they have strong R&D activities that can be joined to ours in cooperative endeavors. Thus, as provided in the Nunn Amendment, we are working with the United Kingdom on an advanced short takeoff and vertical landing program; with the Federal Republic of Germany on an enhanced fighter maneuverability program; and have signed about a dozen final memoranda of understanding with allied partners.

Managing the Defense Technology Base

Military R&D has a long and distinguished history. The role of Archimedes' mirrors in the siege of Syracuse in 212 B.C. has already been mentioned. A more recent example is the 20,000 pounds sterling offered by the British Admiralty in 1714 for the development of a ship's chronometer with a specified accuracy equivalent to establishing a position at sea within ten miles after a six-week journey. The U.S. Army sponsored what was probably its first research project when it launched the Lewis and Clark expedition to explore the Pacific Northwest in 1804–1806. The U.S. Naval Observatory was established in 1842 to improve time keeping for navigation at sea. The Wright brothers responded to Army Signal Corps specification No. 486 of December 23, 1907, for a heavier-than-air flying machine. It was delivered six months later for $25,000.

Complexity — Coordination, Oversight, and Acquisition

Today, these costs and delivery times would be considered small. Defense research is now spread over 73 DoD laboratories, other government laboratories, universities, federal contract R&D centers (FCRDCs), nonprofits, and industrial laboratories throughout the United States, plus some abroad. Its very size has attracted attention, and with that attention came legislation, regulations, specifications, restrictions, special programs, and legal and administrative complexities, creating a whole bureaucracy just to implement, control, oversee, audit, enforce, report, defend, and justify defense research.

About 52 percent of DoD's basic research (budget category 6.1) is carried out in academia, with the balance performed in DoD laboratories (31 percent) and private industry (17 percent). (Nonprofits and FCRDCs account for only a few percent, and are here included under "industry.") Most exploratory development (budget category 6.2) is carried out in DoD laboratories (43 percent). Overall, most of the technology base R&D, including advanced development (6.3A), is carried out in private industry (54 percent), with 32 percent in DoD laboratories and 14 percent in academia.

Programs are administered and managed by the services and defense agencies, and overseen by my organization in the Office of the Secretary of Defense (OSD). In addition to normal face-to-face meetings, symposia, and budget and program formulation activities, the staff conducts a three-tiered series of reviews both to coordinate service and defense agency projects and to gain insight into formulation of an investment strategy. At the staff level, annual science and technology reviews are conducted on a tri-service basis by professional staff members covering specific technology interest areas. About 25 technology reviews are carried out annually in areas as diverse as chemical defense, combat vehicles, aeronautics, aircraft propulsion, and logistics support. At the service level, investment strategy reviews are conducted by my deputy for Research and Advanced Technology, with senior service personnel participation to ascertain program adequacy and to determine directions for the future. Finally, topical reviews are held as required to cover specific and often narrow areas of particular current interest. Topical reviews serve both a coordination and education function. This comprehensive review process works well, but we shall always strive to improve upon it.

DoD Laboratories — Personnel Management and Other Practices

The accomplishments and contributions of the DoD laboratories are widely recognized. For example, they were among the first to discover radar, pioneered in the field of night vision devices, maintained U.S. preeminence in aircraft jet engines, developed the world's quietest submarines, and pioneered in space navigation. The 1986 Nobel Prize for chemistry was awarded to a scientist at the Naval Research Laboratory. These contributions were made by highly qualified, energetic, and motivated scientists and engineers in the laboratories working in a supportive environment fostered by enlightened management. We are concerned about maintaining that quality and motivation in the future, as civil service compensation and employment policies and practices degenerate compared to the private sector. Accordingly, maintaining the quality of the laboratories was a principal issue addressed by the 1987 DSB study on the management of the technology base.

In the fundamentally important area of scientific and engineering personnel, the DSB recommended extending the Navy personnel system demonstration at the Naval Weapons Center at China Lake, CA to all DoD laboratories; increasing the tenure, responsibility, authority, and accountability of laboratory directors to a five-year term; and giving these directors greater procurement and allocation authority. In addition, each service is to select one laboratory to serve as a demonstration project designed to attract and retain highest quality staff, improve contracting effectiveness, improve personnel management, and increase local management authority and accountability. Implementation of these recommendations will require a range of activities, from generating internal policy statements to significant legislation. To spearhead this effort, I am chairing a task force that includes members from the Office of Management and Budget (OMB), the Office of Personnel Management (OPM), and the Office of Science and Technology Policy (OSTP), as well as the Office of the Secretary of Defense (OSD) and the services.

Science and Engineering Education — Universities' Role

American universities are among the best in the world. Their contributions to science and engineering are well known. They also serve as the principal education and training ground for our future scientists and engineers. We are concerned that their capabilities be maintained or improved and also that they attract and graduate an adequate number of Americans. To this end, we have launched a number of initiatives over the past few years, from the Defense University Research Instrumentation Program to prestigious fellowships and university chairs and (starting in 1986) the University Research Initiative (URI). The URI established some 80 Research Centers at universities, each funded at an average of slightly more than $1 million dollars per year. They perform research in areas of interest to DoD, such as microelectronics, materials, and computational fluid dynamics.

Full Industry Participation

No description of DoD–supported R&D would be complete without a mention of two special R&D activities. The purpose of Independent Research and Development (IR&D) is to encourage company-selected and company-sponsored R&D necessary to remain competitive in a technological environment. Most IR&D is done at approximately 250 cost centers run by some 100 major defense contractors. The DoD recognizes IR&D as a necessary cost of doing business and treats it like an overhead charge. A recent independent study (by the RAND Corporation) documented the leverage obtained by DoD for its IR&D expenditures. In most recent years, the authorized "ceiling" for IR&D has been $5.2 billion (which happens to equal the total technology base funding, without SDI). The other special R&D activity is a program designed to involve small R&D companies. It is a government-wide program (most of it funded by DoD) to allow small businesses to undertake smaller R&D efforts. Each Small Business Innovation Research (SBIR) contract consists of a Phase I to show feasibility of a proposal and a Phase II, if Phase I is successful. SBIR contracts have ranged from fundamental research in lasers to the development of portable defribillators. Total funding for SBIR in fiscal year 1988 was almost $200 million.

Conclusion

We cannot anticipate all the issues that will arise in the years to come. Based on our record, we know that we have the capability to deal with such issues, but we cannot rest on our laurels. We must continually adapt to changing needs and circumstances and quickly seize new opportunities. We must be willing to take risks, but in a deliberate way. We must educate and train the best scientists and engineers. We must provide an environment conducive to do good research in our laboratories and motivate our researchers, not place obstacles in their way. We must encourage cooperative research at home and, as much as possible, with our friends abroad. We need to speed technology transition and reduce technology lead time. We need to help make industry fully competitive again and use technology to produce quality products. We must have good communication among our scientists and engineers and take our case to the public and our national leaders.

3

Does Technological Superiority Really Buy Us Security?

Josephine A. Stein

T he military in the United States underwent a profound change with the advent of nuclear weapons. With the ultimate in destructive power at its disposal, the military's orientation was altered from warfighting to deterrence. Deployment of technologically advanced military systems displaced more traditional displays of military force, and weapons became symbols of military prowess rather than fighting machines. But along with the high technology came a series of accidents and failures. What went wrong?

The problem is that overly sophisticated, high-tech military systems simply do not work properly. By placing symbolic technical goals before military or security goals, the Pentagon has run into Murphy's Law: If anything can go wrong, it will.

Some of the consequences are nothing more than comic embarrassments. The president's command aircraft has, for example, managed to jam thousands of garage-door openers.[1] The Bradley Fighting Vehicle, an armored personnel carrier, has repeatedly gotten stuck in the mud.[2] And, at the inaugural demonstration of the Army's highly automated Divad anti-aircraft gun in January 1984, the weapon's computer homed in on the rotating fan of a nearby latrine.[3]

Other accidents are more serious and costly. The explosion of a Titan 34D rocket that reportedly carried the nation's last KH–11 reconnaissance satellite is one such example. The loss of this satellite impairs the ability of the military to carry out vital intelligence-gathering activities.

There has been a series of failures with nuclear weapons guidance systems, including cruise missiles, the Pershing II missile, and the M–X missile. These erratic weapons alarm our allies and our own population, as well as our adversaries.

The common thread in each of these disasters is an urgency to field the latest technology in complex military systems and the inability to stop tinkering with the hardware once it is in the field. The result of decades of "modernization" is an ineffective military infrastructure that has fallen victim to its own fascination with the latest technological developments. Advanced, sophisticated weapons that are prone to failure make the nation less secure, whether those weapons are American or Soviet. The United States may, in fact, be more vulnerable to the Murphy's Law failures of these high-tech military systems than to the diminishing possibility of a deliberate attack by the Soviet Union.

The Problems

Each new weapon or military system sought by the Pentagon is subject to more ambitious performance goals, with greater and greater demands placed on technology. The result is what Mary Kaldor has called a "baroque arsenal,"[4] with weapons technologies becoming progressively more exotic, expensive, and specialized. Each new generation of plane, tank, missile, or ship costs, in constant dollars, an average of two-and-a-half times the previous generation.[5] Cost constraints limit the military to fewer and fewer new weapon systems, so there is a temptation to add more versatility and complexity to each new system. Complexity is, of course, the key to Murphy's Law. The more components, and the more intricate the design, the more things that can go wrong.

Even as the military adds technical complexity to a new weapon design, the pressure mounts to deploy the system as rapidly as possible before the technology becomes outdated. Engineering is extremely difficult under these circumstances. Decisions that are forced through early in the design process impose constraints on the rest of the system, and all too often, they come back to haunt the engineers. Mistakes made early on are costly and difficult to fix with engineering change orders and retrofits. The "fixes" often contain the seeds of future breakdowns that do not appear until the system is in an advanced state of assembly and testing.

Complex systems are broken down into subsystems and components that are subcontracted and sub-subcontracted, diffusing the responsibility for the design and making systems integration more difficult. Coordination of the design and communication between the engineers become more complicated. Management is subject to Murphy's Law just as surely as is engineering.

Keeping to a schedule can be every bit as daunting as the engineering challenge itself. Companies often make optimistic projections on cost and delivery time while competing for contracts, but the Pentagon places the highest priority on keeping to these over-ambitious schedules. Contractors cut corners in quality control and component testing to meet delivery deadlines, so that when the system is assembled for testing, Murphy's Law strikes on the test range.

Political pressures on military contractors and the services all too often put deployment before performance. In one case after another, tests are conducted under the most favorable of conditions; failures are overlooked or papered over. So, it should come as no surprise when so many "operational" systems cannot be counted on to operate as designed.

Our experience with the M–X missile guidance system illustrates several of these problems. For its guidance, the M–X missile uses a complex Inertial Measurement Unit (IMU), a 126-pound sphere the size of a basketball, which houses a hydraulically suspended inertial guidance assembly. The IMU, assembled by Northrop Corporation from 19,401 parts provided by some 553 suppliers, costs about 5.8 million dollars each. Some of the pieces are so small and delicate that they are constructed under a microscope. The M–X is designed to strike within 0.05 nautical miles of its target at least 50 percent of the time. Brigadier General Charles A. May, Jr., head of the Air Force Ballistic Missile Office, boasted that the M–X is "the most accurate missile in the world today."[6]

Perhaps General May is right, if you believe the theoretical calculations. But theory often overestimates the performance of a system in actual use. The "smart" Falcon missile, which was introduced in the Viet Nam War, for example, had a theoretical kill probability of 99 percent; in practice, the Falcon's success record was 7 percent.[7]

The M–X guidance unit has had a long history of problems and delays. Thousands of engineering change orders were placed *after* the M–X guidance unit went into production. Jack Kroll, Northrop's manager of production control on the IMU, told how these design changes made it impossible to duplicate the performance of the IMU prototype.[8]

Eager to deliver the IMUs to the Air Force on schedule, Northrop allegedly lowered some technical standards unilaterally. Some parts went from test station to test station until they passed. Others were not tested at all. Terry Scheilke, a former Northrop employee who was in charge of the company's internal M–X audit, was simply ordered "to close up shop and stop my investigation."[9] On one Air Force inspection, a machine testing an M–X guidance system heat exchanger exploded, slamming a piece of debris into the inspector's foot.[10]

As many as 18 different untested components were installed in M–X missiles to meet a politically motivated deployment schedule. In 1987, the U.S. House Armed Services Committee reported that one-third of the 21 M–X missiles deployed at Warren Air Force Base in Wyoming were not operational. Bryan Hyatt, a former employee of Northrop, told the committee at a hearing, "These missiles stand as much chance of hitting their targets in the Soviet Union as they do of landing here in Washington, DC."[11]

It is one thing for failures to occur in development tests, which are, after all, intended to unmask flaws so that they can be fixed. It is quite another when weapons such as the Divad advance to the system testing stage while the components still do not work properly. It is even more outrageous when, as in the case of the M–X, the systems are deployed.

If the military does produce a complex weapon system that works acceptably well in the hands of its designers, the system may still prove too difficult to operate by soldiers in the field, especially under combat conditions. And even if the Pentagon manages to get a system through with a relatively good performance record in the field with ordinary operators, "upgrades" and modifications continue. It is no wonder that we have had such a string of failures in our military systems; there are so many ways that things can go wrong.[12]

In sum, changing and more demanding performance requirements, pressure to deploy new weapons quickly, fragmentation of the design process, and the difficulty of operating complex systems in the field all create problems. In the end, we are left less secure.

Defense Strategy and Weapons Development

It is not mere coincidence that the performance of so many of our military systems has been so poor. The high-tech failures come as a direct consequence of U.S. strategic doctrine to maintain technological superiority over the Soviet Union. This doctrine is exemplified by articles such as "Winning Through Sophistication: How to Meet the Soviet Military Challenge," in which the authors argue that Soviet advantages in numbers of tanks, aircraft, missiles, ships, and troops must be offset with qualitative superiority in American technology.[13] Technology is seen as a "force multiplier."

Every year, the Joint Chiefs of Staff issue an estimate of our technological standing relative to the Soviet Union.[14] Although the U.S. is portrayed as enjoying a comfortable lead in roughly half of the listed military technologies and deployed systems, the Soviets are judged equal in many areas and even lead in a few, by the Joint Chiefs' reckoning. Arrows are drawn to show where the Soviet Union is closing the gap or pulling ahead. The implication is that unless weapons modernization proceeds according to Pentagon recommendations, the country will be left vulnerable.

According to the strategic doctrine of technological superiority, modernization of the arsenal is an end in itself. "Technological superiority" has come to mean using the latest, highest technology available.

The military keeps abreast of the latest scientific discoveries and technological advancements with the assistance of military contractors and the Pentagon's numerous scientific advisory panels. Military requirements for this technology follow.

In this process, the Pentagon too often confuses novelty and gadgetry with security; it can take a perfectly workable concept and add so many bells and whistles that the system becomes useless. The Aquila Remotely Piloted Vehicle is one such example.

In 1982, the Israelis scored a decisive victory over Soviet-made, Syrian-operated surface-to-air missile (SAM) defenses in Lebanon's Bekaa Valley. The Israelis first sent in small, pilotless aircraft, beaming radar as if they were full-sized aircraft on the attack. When the Syrians turned on their anti-aircraft radar, these signals were picked up by the drones and relayed back to an Israeli command plane. Israeli F–4 Phantom jets set their electronic jammers for the real attack, and fired radar-seeking missiles at the blinded Syrian radars. Once the radars were knocked out, Israeli F–16 bombers could target the SAM batteries at their leisure with conventional bombs. All 19 Soviet-made SAM installations were destroyed without a single Israeli casualty, and the triumphs of Western technology were proclaimed.[15]

The U.S. Army, encouraged by the performance of the little Israeli drone, decided to improve upon the concept with their new Aquila Remotely Piloted Vehicle. Instead of merely smoking out enemy radar, the Aquila would be a highly maneuverable, sophisticated unmanned aircraft that would penetrate enemy territory in search of targets. The Aquila would relay video information back to controllers on a jam-resistant, two-way radio link, so soldiers far behind the lines could maneuver the aircraft. Other data from Aquila would be used to direct conventional artillery and laser-guided munitions to the targets.

The Aquila carried a television camera that could be used to spot targets in the daytime in good weather. Rain, fog, or a smoke-filled battlefield environment were enough to render the TV cameras worthless. An infrared sensor was therefore designed for night-time operations and adverse weather conditions, but a single Aquila could not carry both the optical and the infrared sensors. Optical or infrared units would have to be installed in the field.

The Aquila was first supposed to search out enemy targets and pinpoint their location. There was a laser designator aboard the drone to illuminate targets for laser-guided weapons. At the end of its mission, the Aquila was to return to base, where it could be retrieved in a large vertical net. The Aquila was one of the most complex and demanding battle systems ever attempted by the military.

Not surprisingly, the Aquila ran into serious problems on the test range. The General Accounting Office (GAO), after a comprehensive investigation, identified a number of major problems, including poor launch reliability, difficulty in operating, and frequent breakdowns. Even under the best of conditions, the television image quality was poor. Only 22.7 percent of moving targets and 12.3 percent of stationary targets were detected out of the hundreds that were placed in Aquila's field of view in repeated tests. These percentages dropped substantially when the targets were camouflaged. One operator complained that using Aquila to hunt for targets was like peering through a drinking straw.[16]

Unrealistic test conditions led GAO to question whether the Aquila could operate in the presence of an enemy, but perhaps GAO was being generous. Even without an enemy, the Aquila completed its test course successfully only seven times out of a total of 105 trials.[17] A telling observation made by GAO was that the system was too complex for its operators. Aquila was an invitation for the consequences of Murphy's Law.

When Defense Secretary Frank C. Carlucci directed the military services to cut $33 billion from their proposed $332 billion budget for fiscal year 1989, the Army finally threw in the towel and canceled the Aquila.[18]

Although the more sophisticated Aquila failed, there was nothing wrong with the original, far simpler Israeli technology. In fact, in July 1985, when John F. Lehman, Jr., then Secretary of the Navy, ordered the rapid procurement of a new, unmanned surveillance aircraft, the Navy placed its order with a firm called Mazlat, an Israeli company.[19]

In the U.S. quest for technological superiority, the excuse is often given that we seek an advantage over the Soviet Union only for the purpose of negotiating arms control agreements. Somehow, all the bargaining chips tend to get deployed. The desire to deploy an advantage, no matter how illusory or transitory, has, in some cases, made us more vulnerable. The Pentagon often forgets that a weapon that is more threatening to the Soviet Union will be more threatening to the United States once the Soviets deploy a comparable system.

For example, in the case of the multiple independently-targeted re-entry vehicle (MIRV), the security consequences of having such weapons deployed by both superpowers were not thought through *a priori*. The Soviets were anxious to prevent MIRV deployments when the United States first developed the capability. Conscious only of gaining what the United States saw as a decisive advantage that analysts said the Soviets could not match for decades, the Pentagon went ahead and deployed MIRVed missiles in 1970.

About five years after the Soviets deployed MIRVed systems on their heavy booster rockets, the United States noticed a window of vulnerability. By multiplying the number of nuclear warheads on the most accurate Soviet delivery systems, a single Soviet missile would in theory be able to destroy several U.S. silo-based intercontinental ballistic missiles (ICBMs), each armed with several nuclear warheads. Even some of the most ardent proponents of deploying MIRV have come to realize that we would have been more secure without MIRVs on either side.

To compensate for the new threat from Soviet MIRVs, the Pentagon proposed a new ICBM, which, due to its basing scheme, was to be less vulnerable than the fixed, land-based Minuteman missiles. This new, more accurate missile, which became known as the M–X, was designed to be just slightly larger in diameter than Minuteman silos. At least 30 basing modes for the M–X were considered before the decision was made to place them in Minuteman missile silos, which had to be specially enlarged. Research continues on rail-mobile M–X basing, but it faces stiff political opposition.

The United States could easily have stuck with its Minuteman force, once one of the primary original justifications for the M–X — more secure basing — had vanished. If the old Minuteman rockets had suffered deterioration with age, the

U.S. could simply build new Minutemen of the same design. Even though the M–X with its ten warheads would be a more inviting target than the Minuteman with its three, the M–X is also bigger, newer, and more accurate than the Minuteman. So, in keeping with the doctrine of military modernization for its own sake, the M–X entered the arsenal.

The M–X was sold to the American people as a bargaining chip to get the Soviet Union to reduce its land-based ICBM forces. Today, the M–X missiles are deployed, and there's little hope left that Reagan Administration negotiators will produce a strategic arms agreement with the Soviets. The M–X has not solved the window of vulnerability, nor has it improved stability or national security.

Another reason the military ends up with costly, ineffective weapons has to do with parochial interests within Congress and the Pentagon. Nothing illustrates this better than the case of the B–1 bomber, an advanced plane with adjustable wings and a sophisticated electronic warfare system designed to jam multiple Soviet air defense radars. The need for the B–1, which is designed to penetrate Soviet air space, has always been controversial.

Even President Carter could not cancel the "Flying Edsel," as it is called, despite repeated attempts during his administration. Rockwell International, the prime contractor, has its production facility in Palmdale, California, in "B–1 Bob" Dornan's congressional district and in senator Alan Cranston's state. Both the congressman and the senator fought successfully to keep B–1 jobs for their constituents, with the assistance of many of their colleagues. The B–1, as it turns out, is produced by some 5200 subcontractors in 48 states.[20]

What has this produced? The B–1 has had software problems in its flight-control system. A computer that is supposed to spot malfunctions sets off false alarms.[21] There are bugs in the B–1's terrain-following radar, which jams its own signals.[22] The B–1, with all of its advanced avionics and electronic sophistication, broke down at the 1987 Paris Air Show because of an electrical failure. One of the planes was brought down by a collision with a large bird, killing three crew members.[23] Continual problems prompted Les Aspin, Chairman of the House Armed Services Committee, to say in exasperation, "The U.S. Air Force has been a greater threat to the success of the B–1 bomber than the Soviet Union."[24]

Finally, in July 1988, the Air Force announced, "The existing system was discovered to be incapable of meeting the ...specification requirements."[25] This means that the electronic warfare system, the centerpiece and raison d'être of the B–1, has such serious design flaws in its basic architecture that the Air Force may be forced into developing a whole new system. Despite protracted testing while the plane was under development, these design flaws were only identified after the plane was declared operational. No one knows how much it will cost to correct the problem.

Parochialism also haunts the halls of the Pentagon. In peacetime, the military services focus their competitive instincts on each other instead of on the Soviets. This leads to duplication of effort in weapons research and development and procurement. The prestige of modern weaponry leads to an internal arms race between the services for the latest, most advanced hardware. We have ground-launched, air-launched, and sea-based cruise missiles, for little apparent reason except that we have an Army, Air Force, and Navy.

Richard L. Garwin, an IBM research fellow who consults for the Pentagon, once observed, "The biggest game in town is the U.S. Navy vs. the U.S. Air Force, not the U.S. vs. the Soviets."[26] If the Air Force gets a new Stealth Advanced Tactical Fighter, then the Navy has an Advanced Tactical Aircraft. It took Congress years to notice that the planes were virtually identical in some important respects. Now that the planes are nearing the production phase, Congress has finally demanded that the Air Force and the Navy undertake joint development of the two aircraft.[27]

In peacetime, military contractors also turn their competitive instincts toward each other and the Pentagon instead of the Soviets. James Fallows has described how military firms' sales pitches, designed to attract the lucrative Pentagon contract, evolve. First comes the "Vegematic Promise" (It slices, it dices, and does just about anything under the sun.) Next comes "the rosy prospect," which is based on optimistic analyses of timelines and costs. The third stage is "the technical leap," in which the contractor assures the Pentagon that unsolved technical challenges will be met. Next is "the unpleasant surprise," followed by the final "collapse of the house cards."[28]

But since victory in the battle for the military contract is so important to the survival of the firm, Pentagon contractors have gone beyond simple misleading advertising. The "Ill-Wind" Pentagon procurement scandal is only symptomatic of the degree to which the military procurement process has become corrupted. Companies make promises they know they cannot keep, and the military passes the promises along to Congress in order to get funding. The country is left with unworkable weapons systems.

The United States may look foolish when the military shoots itself in the foot with M–X missile test equipment debris. But the poor performance of so many high-tech military systems is not just a matter of lost face and wasted money. Even in peacetime, sudden and unexpected failures of early warning systems can set off military alerts that are presumably monitored with some alarm by the electronic intelligence network of the other superpower.

On October 3, 1979, the North American Air Defense Command (NORAD) observed a piece of space debris re-entering the atmosphere and mistook it for a missile attack.[29] In November of 1979, a computer wargame tape was loaded into a computer, and for six minutes, NORAD thought the U.S. was under attack by Soviet missiles. Ten planes at three different air bases took off before the mistake was recognized. A false alert from the World-Wide Military Command and Control System (WWMCCS) in early June 1980 sent Air Force personnel scrambling for their planes. A congressional investigation discovered that in an 18-month period ending June 30, 1980, there were 3703 false warnings of nuclear attack, of which 151 were categorized as "serious."[30]

Each of these false alarms was a unilateral failure on the part of the complex network of sensors, computers, and individual military personnel which comprise the U.S. early warning system. In the nuclear age, this system is perhaps the single most important element of national security on a day-to-day basis. Presumably, the Soviets have experienced similar false alarms. That we have managed to avoid catastrophe so far is not very reassuring.

Mistakes sometimes assume tragic proportions when the military misdiagnoses an apparent threat. Most failures of complex human/machine systems result from an unforeseen combination of events, which becomes compounded by errors of judgment. For the military, this sort of situation is exacerbated by the stresses of combat. There is no better illustration than the tragic failures of U.S. naval defense systems in the Persian Gulf.

The U.S.S. *Stark*, stationed in the tense waters of the Persian Gulf in May 1987, was equipped with modern radars, sophisticated electronic equipment, and computer-controlled Phalanx gun defenses. Nevertheless, it was a human lookout who first spotted the missile fired from an Iraqi Mirage F–1 warplane as it approached the ship, seconds before the missile hit. According to the skipper of the *Stark*, Capt. Glenn R. Brindel, there was no time to activate the Phalanx defense system.[31] Thirty-seven sailors died in the attack.

In July 1988, the U.S.S. *Vincennes*, equipped with the ultrasophisticated, $1 billion Aegis naval combat system, mistook a civilian Airbus A300 for an attacking F–14 fighter and fired. Two hundred and ninety people aboard the regularly scheduled Iran Air Flight 655 to Dubai died. There was a great deal of confusion and many contradictory stories about the performance of the Aegis system over the following months. The Navy claimed that the plane was traveling at 450 knots (about 520 mph), at an altitude between 7,000 and 9,000 feet, and that the plane was descending in an unambiguous attack profile. Other observers put the altitude of the plane at 12,000 feet — and ascending — and noted that the Airbus is not even designed to fly as fast as 450 knots at 9,000 feet.[32]

The Aegis naval combat system is a highly automated system designed to track and identify hundreds of targets in three dimensions, display the information on large video screens, and engage in many simultaneous, computer-controlled attacks. *Washington Post* writer Philip J. Hilts wrote, "The Aegis system aboard the *Vincennes* is as high tech as high tech gets in modern warfare."[33]

The case of the *Vincennes* is a classic example of how things can go wrong when complex military systems are placed at a high state of alert. Minutes before the ill-fated Airbus took off, there was an exchange of fire between Iranian gunboats attacking a U.S. helicopter and the U.S. Navy ships *Vincennes* and *Elmer Montgomery*. The *Vincennes* picked up transmissions from a military aircraft. As data from the approaching airliner came in, the *Vincennes* was executing a maneuver and, in the confusion, the decision was made to fire before visual contact was made. Despite the fancy gadgetry aboard the *Vincennes*, there is still no way that its radar can determine the size or shape of an object in its field of view, let alone identify it reliably as friend or foe.

The Aegis combat system was controversial long before the *Vincennes* made its lethal mistake. Congressman Denny Smith, a Republican from Oregon, became suspicious in 1983 when the Navy reported that the Aegis cruiser *Ticonderoga* achieved a perfect score in test firings against thirteen target planes in a simulated attack. It was later reported that the *Ticonderoga* had, in fact, hit only five of twenty-one targets, which Smith, after investigation, said was more accurate. The Navy "improved" the performance of the system in the next round of tests, according to Smith, by making the tests easier.[34]

Whether the *Vincennes* accidentally attacked the Iranian Airbus because of technical failures in the Aegis, human error, or some combination is not known at this writing. It may turn out that the *Vincennes* did, in fact, pick up signals from a F–14 in the area, just as the Soviet Air Force, in an earlier tragedy, may have picked up signals from a U.S. RC–135 spy plane that was in the vicinity at the time the civilian Korean aircraft KAL 007 was shot down. On combat alert, military commanders understandably feel that when confronted with ambiguous data, it is better to fire first at the apparent threat and deal with the consequences later.

In a situation where nuclear weapons are involved, the consequences of a mistake can be catastrophic. In a serious confrontation or conflict between the superpowers, plans for predelegation of nuclear attack authority exist, although they are heavily classified.[35] Pitting the instincts for self-preservation on the battlefield, in a nuclear-armed analog of the Korean airliner or *Vincennes* incidents, against the consequences of a mistaken nuclear attack is almost too awful to contemplate.

Both the United States and the Soviet Union are threatened by mishaps, errors, accidents, and misjudgments, as long as complex, dangerous military systems remain deployed on either or both sides. The "hot line" connecting Washington and Moscow and joint crisis management centers are a measure of protection, but they are also one more element in the complex, coupled military forces of the superpowers.

Technology and Security

There are steps that can be taken to reduce the threat posed by oversophisticated, complex weapon systems. The first is to *challenge the assumption that higher technology makes us safer*. It is time that we face the inescapable truth that Murphy was right. This means a return to a few, basic engineering principles, the most important of which is *make it simple*. It is no coincidence that some of our best and most reliable military systems — the Sidewinder Missile, the U–2 spy plane, the SR–71 Blackbird reconnaissance plane, and the Ada programming language — are products of individual designers and the engineering design teams under their direct supervision

In almost all cases, making it simple will also *make it cheaper*. For military industry, this may not be good news, but we cannot afford to let the prospects of increased profits from the sales of more complicated and advanced technology weapons distort our national security priorities. Military procurement reform applies not only to Pentagon contractors, but also to the political environment in which military contracts are awarded and congressional decisions are made. Pork barrel military contracting is thoroughly inappropriate when our national security is at stake.

Another engineering maxim is *use the right tool for the job*. This means assessing the military need before developing the weapon, not succumbing to the technology salesman and scrambling for a justification after the fact.

There is one other important step that must be taken to reduce the risks of an overgrown, overly complex military machine. We need to *reassess the job for which the right tools are needed*. The reality is that there is very little military threat to the United States, surrounded as we are by Canada, Mexico, and two oceans. Rather than defending our borders and our shores, which seems to be left to the Immigration Service and the Coast Guard, our military is preoccupied with maintaining a massive strategic arsenal of baroque, high-tech weaponry and with projecting force from the Philippines to the Persian Gulf. The Pentagon has replaced cold assessments of realistic military threats to the United States with high-tech posturing toward the Soviet Union and intimidation toward the Third World.

It may be counterintuitive to centuries of military tradition, but our nation would be much more secure if we and the Soviets both backed off the threatening postures and the technological sophistication that underlies those postures. The size, diversity, and complexity of the superpowers' nuclear arsenals have become a security liability. Likewise, placing hundreds of thousands of American and Soviet combat troops, armed with complex and dangerous military equipment, in or near foreign countries is provocative and dangerous.

Recent high-level military discussions between the Chairman of the Joint Chiefs of Staff, Admiral Crowe, and his Soviet counterpart, Marshal Akhromeyev, have been a very positive development toward building a common understanding of security between the United States and the Soviet Union. It is important to remember that although rivalry between the superpowers has dominated the political scene for decades, our two nations were allies in World War II. By building on our historic ties and joining in further discussions in a common quest for mutual security, the U.S. and the Soviets can reassess military needs and configure their military forces accordingly.

The political and financial costs of maintaining hundreds of foreign military bases make less and less sense. As economic pressures rise in both the United States and the Soviet Union, the need for supporting global military infrastructures will be challenged. We face a historic opportunity to pull back from the brink of disaster the world over.

It is time that we recognize the dangers of excessive deployments of Murphy-prone military systems. We can stop shooting ourselves in the foot with overly

complex, sophisticated, high-tech military systems, and get down to the challenge
of placing national and international security first.

Notes

1. Peter G. Newmann, "Illustrative Risks to the Public in the Use of Computer Systems
   and Related Technology," ACM Committee on Computers and Public Policy, April
   5, 1988.

2. Steve Johnson, "2 Bradley Fighting Vehicles Get Stuck in River-Crossing Test,", *San
   Jose Mercury News*, December 11, 1987.

3. Hedrick Smith, *The Power Game: How Washington Works*, Random House, New York,
   1988.

4. Mary Kaldor, *The Baroque Arsenal*, Hill and Wang, New York, 1981.

5. James Fallows, *National Defense*, Random House, New York, 1981.

6. Molly Moore, "M–X Reliability in Question," *The Washington Post*, December 28,
   1987.

7. Seymour Melman, *Profits Without Production*, University of Pennsylvania Press, 1983.

8. James Kitfield, "Unguided Missiles," *Military Forum*, April 1988.

9. *Ibid*.

10. Molly Moore, "M–X Reliability in Question," *The Washington Post*, December 28,
    1987. (Moore, *op. cit.*)

11. Trish Gilmartin, "Northrop Mismanagement Led to Defective MX Guidance Systems,
    Congress Told," *Defense News*, June 15, 1987.

12. The Soviets, too, have suffered from accidents with high-tech military systems, al-
    though little is known about them. Tales are told of a new submarine lost at sea, kill-
    ing all the sailors aboard. Soviet leader Khrushchev reportedly told President Nixon
    that the engine cutoff system of a Soviet ballistic missile once failed during a test, send-
    ing the missile on a course to Alaska. Fortunately, the missile fell into the sea. John
    Hanrahan, "Whose Finger's On the Button?" *Common Cause Magazine*, Vol. 10, No.
    1, January/February 1984.

13. William J. Perry and Cynthia A. Roberts, "Winning Through Sophistication: How
    to Meet the Soviet Military Challenge," *Technology Review*, July 1982.

14. *United States Military Posture*, prepared by the Organization of the Joint Chiefs of
    Staff, United States Department of Defense, annual statement.

15. "Killer Electronic Weaponry," *Business Week*, September 20, 1982.

16. "Aquila Remotely Piloted Vehicle: Its Potential Battlefield Contribution Still in
    Doubt," *United States General Accounting Office Report*, GAO/NSIAD-88-19, October
    1987.

17. *Ibid*.

18. Molly Moore, "Defense Department in Fiscal Retreat: 'Where Are We Headed?'" *The
    Washington Post*, January 6, 1988.

19. Joe Pichirallo, 'Extraordinary' Process Used to Buy Navy Drone," *The Washington Post*, July 28, 1988.

20. Nick Kotz, "Money, Politics, and the B–1 Bomber," *Technology Review*, April 1988.

21. Timothy Noah, "It's Hard to Flunk a Weapons Test," *Newsweek*, September 7, 1987.

22. Peter G. Newmann, "Illustrative Risks to the Public in the Use of Computer Systems and Related Technology," ACM Committee on Computers and Public Policy, April 5, 1988.

23. Molly Moore, "Air Force Blames Pelican in Fatal Crash of B1B," *The Washington Post*, January 21, 1988.

24. *Sane World*, Committee for a Sane Nuclear Policy, Washington, DC, Summer 1987.

25. Molly Moore, "New Flaws Found in B–1 Bomber," *The Washington Post*, July 10, 1988.

26. "Guns vs. Butter," *Business Week*, special report, November 29, 1982.

27. "Congress Orders Joint Development of ATA, ATF," *Aviation Week and Space Technology*, July 18, 1988.

28. James Fallows, *The New York Review of Books,* December 18, 1986.

29. Daniel Ford, *The Button: The Pentagon's Command and Control System, Does It Work?* New York: Simon and Schuster, 1985.

30. Edward Dolnick, "Can Computers Cope With War?" *The Boston Globe*, December 10, 1984.

31. William Matthews, "Stark Incident Raises Questions About Wartime Survivability of U.S. Frigates," *Defense News*, May 25, 1987.

32. "Incidents in the Persian Gulf," *Congressional Quarterly*, July 9, 1988.

33. Philip J. Hilts, "Aegis Subject to Human Error," *The Washington Post*, July 6, 1988.

34. Hedrick Smith, *The Power Game*, New York: Ramdom House, 1988.

35. John Hanrahan, "Whose Finger's On the Button?" *Common Cause Magazine*, Vol. 10, No. 1, January/February 1988.

4

Relating Military R&D to the Civilian Economy: Policy, Perceptions, and Paradigms*

Alexander H. Flax

For most of the years since World War II, proponents for various points of view concerning the merits of defense spending in general, and defense research and development (R&D) spending in particular, have debated over the significance of the collateral benefits of defense R&D to the civilian sector (variously characterized as either "spin-off" or "trickle-down," depending on the point of view of the speaker.)[1] Some have argued that spin-offs from defense R&D have had a major influence in spawning new industries and giving the United States substantial advantages in international competition for civilian goods. Others have maintained that relative to the large resources expended on defense R&D, the contribution to civilian industry is negligible and perhaps in some ways even negative. Since the United States spends well over half of its R&D budget on defense (see Table 1), larger than many of its allies, this is an important consideration.

For most of the years since World War II, almost no one in the U.S. defense establishment planned R&D with commercial applications in view, and the DoD was neither required to nor inclined to justify its R&D programs on the basis of expected spin-offs to the civil sector.

However, the temper of the times has been changing in the United States. With the rising concern about U.S. competitiveness in domestic and global markets,

*This article is reprinted with the permission of Kluwer Academic Publishers, Dordrecht, Holland, and is based on a paper presented at the Conference on the Relationship Between Defense and Civil Technologies, Wiston House, Sussex, England, 21–25 September 1987.

**Table 1.   Distribution of Government R&D Expenditures
by Objectives, 1984**

	U.S.	France	FRG	Japan	U.K.
Agriculture, forestry, fishing	2.1	4.7	2.4	10.9	5.0
Industrial development	0.2	11.7	11.6	6.1	8.5
Energy	5.8	7.9	15.0	14.0	4.8
Transport & telecommunications	2.5	2.2	1.1	1.4	0.5
Urban & rural planning	0.1	1.3	1.1	1.1	1.0
Environmental protection	0.5	0.5	2.8	1.4	1.2
Health	11.3	3.8	3.2	2.5	3.5
Social development	1.2	1.4	2.4	0.7	0.7
Earth & atmosphere	1.4	2.0	2.2	1.1	1.7
Advancement of knowledge	5.2	26.5	44.4	53.5	19.6
Advancement of research	3.9	16.2	11.4	1.7	5.1
General university funds	0.0	10.3	33.0	51.8	14.6
Civil space	5.2	5.8	3.9	4.4	2.2
Defense	66.0	31.3	9.8	2.8	49.4
TOTAL (%)	100.0	100.0	100.0	100.0	100.0

Note: Less than 1 percent of French funds and 1.8 percent of U.K. funds were not elsewhere classified. The "Advancement of knowledge" category should not be equated with basic research. U.S. figures are for the Federal Government which does not provide general university R&D funds.

Source: International Science and Technology Data Update 1986. Report prepared by Directorate for Scientific, Technological, and International Affairs, Division of Science Resource Studies. National Science Foundation.

technology transfer to the civilian sector from all federal programs has become an idea whose time has come. As Table 2 shows, for the amount of investment the United States makes in defense R&D, the relative return in terms of exportable technologies is disproportionately low.

The Stevenson–Wydler Act of 1980 and its later amendments attempted to establish information transfer and other relationships between federal laboratories and industry. At the same time, there was rising concern over the burgeoning Reagan Administration defense budgets. These factors, and perhaps others, brought about a significant change in the DoD's rhetoric on R&D spin-offs. Although the spin-offs from DoD's R&D programs are not a primary objective of those programs, spin-offs have taken a place in the supporting arguments to help justify the DoD R&D budget. Thus, in presenting the budget for basic research programs, the DoD pointed to past payoffs of these programs in lasers, computerized systems, communications, and advances in medical services.[2]

The Strategic Defense Intiative (SDI) program, which, as originally planned in the Fletcher Study, gave no consideration to potential civilian benefits, is now being

Table 2. R&D and Technology-Intensive Export Performance

	U.S.	Japan	FRG	France	U.K.	Sweden
R&D expenditures as % of GNP	2.8	2.8	2.7	2.4	2.2	2.5
Nondefense R&D as % of GNP	1.9	2.8	2.6	1.9	1.5	2.3
Government R&D for defense & space (%)	74	8	17	37	54	29
R&D scientists & engineers per 10,000 in labor force	69	63.2	49.1	41.2	32.8	39.0
Share of technology-intensive exports* (%)	24	19	15	8	9	2
Share of technology-intensive exports per million in the labor force	.20	.31	.54	.34	.32	.48

*Reflects information from 24 reporting countries on exports to and from each of nearly 200 partner countries. Technology-intensive products are defined as those for which R&D expenditures exceed 2.36 percent of value added.

Source: Lederman, L.L., "Science and Technology Policies and Priorities: A Comparative Analysis." *Science,* September 4, 1987. Tables I and II, pp. 1125-1133.

portrayed as a rich source of promising spin-offs, thus echoing in some degree many European perceptions of the program.[3] Of course, given the wide spectrum of science and technology in the DOD R&D program or in the SDI program, it is not surprising that such potential readily can be found. How effective technology transfer through spin-off really is, and how much technology *per se* has to do with international competitiveness and successful commercialization of the resulting products in the civilian economy, are the main issues in the ongoing debate in the United States. In this paper, we will attempt to address these issues and the related questions of how to define in a meaningful way some of the terms entering the debate. The focus in this paper is primarily on military R&D, not total military spending. Total military spending obviously can have large effects on the civilian economy, but not necessarily through direct effects on civilian technology that are specifically related to military R&D.

The Composition of Military R&D

Comparisons of federal government-supported military and civilian R&D in the United States are not really meaningful. Prevailing economic and political doctrines in the United States have, until fairly recently, distinctly been based on the premise that technical development or improvement of commercial manufactured products, except for reasons of public health and safety, is no business of the government. There are a few notable exceptions — for example, aerospace and nuclear energy — but these have been justified for reasons closely related to defense interests and concerns.

Table 3. Defense Department RDT&E Budgets (in thousands of dollars)

Summary of program by activities:	1986 actual	1987 est.	1988 est.	1989 est.
Technology base	3,232,381	3,233,273	3,420,639	3,643,323
Advanced technology development	4,066,807	4,930,142	7,163,170	8,477,821
Strategic programs	7,508,952	8,124,680	9,989,998	9,222,921
Tactical programs	10,265,746	10,998,112	13,726,645	13,826,516
Intelligence and communications	4,525,113	4,922,771	5,262,070	4,841,956
Defensewide mission support	4,077,044	4,043,005	4,156,415	4,190,147
TOTAL DIRECT	**33,676,043**	**36,251,983**	**43,718,937**	**44,202,684**

Source: Office of Management and Budget, Budget of the U.S. Government, Appendix, Fiscal year 1988

The distribution of Defense Department research, development, testing, and evaluation (RDT&E) is shown in Table 3. This shows that about 10 percent of the defense fiscal year 1986 RTD&E budget is spent on the technology base (basic and applied research) — the 6.1 and 6.2 categories in the defense programming system. Another 12 percent was spent on advanced development (6.3A in the program categories) that includes prototypes and demonstrations of systems and subsystems not yet validated or approved for development for production. (Over half of the latter category is for SDI). The strategic, tactical, communications, and intelligence items amounting to 66 percent of the budget support the development and product improvement of specific systems and subsystems for the operational military inventory. The last item, defense-wide support, covers test ranges, facilities, and other activities necessary to carry out the R&D programs on a whole.

The budget for development activities is a consequence of decisions to develop particular weapon systems and subsystems for the government's own use and, in most instances, the government itself is the only prospective buyer. The annual sums required to pay for specific major systems developments are often of the same order of magnitude as the entire technology base program.[4] For example, in the 1984 fiscal year, funding for the M–X and Trident missile programs alone exceeded the entire technology base program. Such large development expenditures have little, if anything, to do with policy determinations as to the appropriate level of military R&D, much less with any linkage to federally supported civilian R&D. Military development expenditures are directly related to force structure and modernization policies. The fiscal year 1988 development program budget includes the small mobile ICBM, the advanced technology bomber, the Trident II missile, the light helicopter (LHX), a new heavy military transport aircraft (C–17), a new nuclear attack submarine, and an advanced tactical fighter, as well as product improvements for existing systems such as the A–6, F–14D, F–15 aircraft, and the M1A1 tank.[5]

It is probably not reasonable to expect much direct payoff to the civilian sector from the major system and subsystem development programs of DoD, although secondary benefits may accrue from raising the general level of industrial technology in such fields as gas turbine engines, high-strength composite structures and materials, avionics, machine tool and manufacturing systems, inspection techniques, and instruments. But almost all of these benefits are two-way — progress in civilian industry often benefits military technology as much or more than the converse. Moreover, the benefits of interchange are highly variable across industrial sectors. Thus, the value of technological interchange with military R&D is very much greater in aircraft, aircraft engines, satellites, and space launch vehicles than in the automotive industry, consumer electronics, and electrical appliances.

These latter industries — along with chemicals, pharmaceuticals, and textiles and fibers, which are even more remote from military technology — are major components of the civilian sector of industry that depend on technological progress to maintain their competitiveness in international markets. Contributions from DoD R&D to these and similar industries, if they occur, must traverse indirect routes which are difficult to trace. No satisfactory analytical measure for such transfer seems to exist.

Science and Technology and Civilian Industry

In this century, there have been some major advances in basic science which took place in the laboratories of a few very large industrial corporations. The most relevant of these for high-technology industry was undoubtedly the invention of the transistor at the Bell Telephone Laboratories in the United States. (This invention, because of the workings of U.S. antitrust laws, was not exploited entirely for company advantage. Rather, licenses were freely granted to all comers, foreign and domestic, for a nominal fee.) However, in the United States, by far the overwhelming share of basic research is performed in universities, and this research is freely available to applied researchers and industry in every country. This availability of basic research results means that there may be little, if any, direct comparative industrial advantage to the nations that support basic research. This research enriches equally all nations that are prepared to receive it — that is, virtually all of the highly developed industrial nations.

Studies made in the United States show some amazing statistical correlations between a nation's success in basic research in physical science and economic and industrial growth, as indicated by Nobel prizes in physics and chemistry per capita and similar measures of basic research output and quality.[6] As Figures 1 and 2 show, rates of economic growth and growth in manufacturing productivity are strongly negatively correlated with the number of Nobel prizes per capita. Obviously, if basic science feeds technology in support of national economic growth and manufacturing productivity, the process is determined as much or more at the receiving end as at the transmitting end. This is apparently also true for basic or generic technology, although it is difficult to find so simple (and perhaps simplistic) a measure as Nobel prizes per capita.

Figure 1. Growth in Gross Domestic Product and Nobel Prize-Winning in Physics and Chemistry

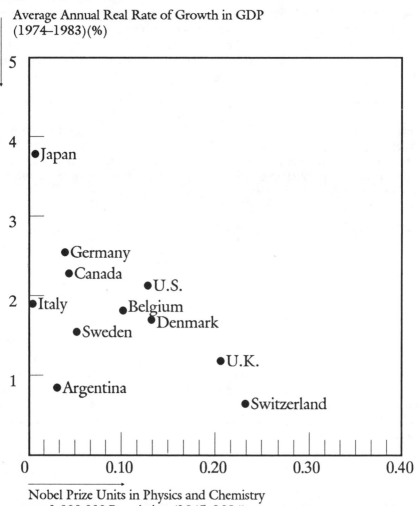

Average Annual Real Rate of Growth in GDP
(1974–1983)(%)

Nobel Prize Units in Physics and Chemistry
per 1,000,000 Population (1945–1984)

Figure 2. Growth in Manufacturing Labor Productivity and Nobel Prize-Winning in Physics and Chemistry

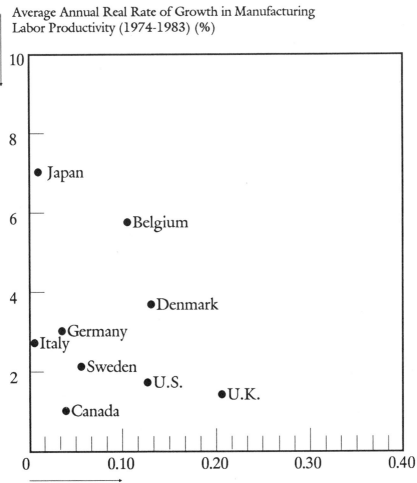

Average Annual Real Rate of Growth in Manufacturing
Labor Productivity (1974-1983) (%)

Nobel Prize Units in Physics and Chemistry
per 1,000,000 Population (1945–1984)

The recent extensive study and debate in the United States over the issues of competitiveness has led to a deeper understanding of what is involved in converting science and technology into commercially successful products for the global marketplace. It has become quite clear that science and technology by themselves cannot, in today's global economy, be counted upon to lead to success in the civilian economy. Many macroeconomic and microeconomic factors and the government policies influencing them play at least equal, and sometimes greater, roles. Another major factor is the structure of industry for introduction of innovative products. On the one hand are relatively few giant, vertically integrated, horizontally diversified Japanese corporations; on the other hand are the myriad small companies that start up supported by speculative venture capital in the United States.

The Japanese,[7] while seeking also to achieve the innovative advantages of small U.S. start-up companies, are quick to point to the overwhelming advantage of their giant corporations in mustering the financial, human, and organizational resources necessary for major assaults on the world's market and, perhaps even more important, in providing the long-term staying power necessary to fully exploit a new technology over its life cycle in the marketplace through continuous product and process improvements and expansion of applications. The spokesmen for Japan also point out that many of the venture capital-supported high-technology companies in the United States are usually singularly lacking in staying power in the marketplace and are thus easily bested in international competition over the long term.

Thus, we are now on notice that we cannot measure the impact of military R&D on U.S. civilian industry by looking only at the success, or lack thereof, of U.S. products in the civilian marketplace. It is obviously highly questionable to attempt to relate the spectacular decline in sales of U.S.-designed and -manufactured consumer electronics in world markets solely to the fact that U.S. R&D was inferior, whether or not that inferiority was brought about by the military drain on intellectual resources in electronics or because of the bad examples in the ordering of priorities (for example, cost versus performance) design and manufacturing engineering set for U.S. industry by military contract experiences. These arguments, which have been advanced for over 30 years in the United States, are now seen to be, at best, vast oversimplifications. Even if there had been no military R&D in electronics in the United States since World War II, and if somehow the same electronic science and technology had evolved from the civilian sector or been transferred from other countries, there is little evidence that U.S. industry would be any further ahead in today's marketplaces for consumer electronics. However, as will be discussed, military procurement may be at least as important as military R&D in its effect on American industry.

The decline of the U.S. competitive position in industries based on older and less rapidly changing technologies (such as the automotive industry) certainly supports such a conclusion. The direct influence of either military R&D or military procurement on the U.S. automotive industry since World War II has surely been minor. (Indirect effects through products of other industries such as computers, electronics and composite materials are not the issue here.) Yet, over the past thirty years foreign designed or manufactured automobiles and components have acquired 30 percent of the U.S. market and would have cornered even more of it were it not for "voluntary" restraints on total automobile imports from Japan. Even the recent 40 percent drop in the value of the dollar versus the yen and many European currencies has not seemed to do much yet to reverse the trend of declining shares for American-designed and -built automobiles. By contrast, two areas which have received considerable defense R&D and procurement support over the past 25 years are aerospace and computers, and they are in reasonably good shape in domestic and foreign markets. In fact, they are among the few manufacturing sectors where the United States still retains a positive balance of trade. This is not to argue that defense R&D is the primary cause of these happy circumstances, but only that the interactions here must be carefully examined.

The more complex view of the factors which influence industrial success in the commercial marketplace, which has become more apparent recently, makes the separation of the effects of technological factors much more difficult. If U.S. firms in some industrial sector are not faring well in world competition, how much can we attribute to the failure of our large investments in military R&D to provide them with useful transferred technology? What objective measure can we use to determine whether more R&D investment specifically oriented to civilian products would reduce or eliminate their problems in competing in domestic and foreign markets?

R&D and "Product Realization" in the Civilian Economy

Over the past decade, intensive study has been given in the United States to the causes for the decline in the competitive position of an increasing number of U.S.-manufactured products in both domestic and international markets. This decline has been marked not only by an unfavorable trade balance of major proportions, but by the near complete disappearance of U.S. production of many items, especially in consumer electronics. Many economists of various schools have propounded theoretical explanations of the observed situation from macro- and microeconomic points of view and have characterized its significance in terms ranging from "normal" to "alarming" depending on their particular economic, social, and political predilections and outlooks.

Yet, many in U.S. industry who have studied the differences between their operations and those of their more successful competitors abroad have discovered that differences in macroeconomic factors and trade policies are not by themselves capable of explaining their loss of markets. Rather, they have perceived that their more successful competitors have succeeded because they could, at the same or lower costs, manufacture products of higher quality, reliability, and performance which were better suited to the needs of the buyers. Moreover, to achieve these results, it was necessary to treat all aspects of their operations as an integrated system without arbitrary and artificial compartmentalization of such functions as research, design, development, manufacturing, quality control, distribution, marketing, and product servicing.

In this systems approach, the management and utilization of human resources forms a most essential element; the changing nature of societies and cultures and of the workforces requires careful and understanding attention. The workforce management concepts underlying Henry Ford's assembly lines and Frederick Taylor's minimization of time, motion, and energy expenditure of the operatives in plants are no longer adequate to deal with today's products and today's labor forces. Thus, the technology of modern manufacturing is as much the technology of the processes of and functions required to bring a product from concept and design to the ultimate user as it is the technology to be found physically embedded in the product. At the American Telephone and Telegraph Company, they have coined a name for this systems approach — they call it the "product realization process."

What can be said about R&D in the product realization process? First, we can say that R&D is important but second. We can also say that, in its traditional forms, it is totally inadequate to assure the success of a product in the commercial marketplace. If a technology is changing rapidly enough so that a rapid-fire succession of products with or having essentially new and different attributes can be introduced into the marketplace and, if one firm (or one nation's firms) is so far in front of this changing technology that competitors cannot keep up with the rapidly advancing front of technology over times shorter than the time interval between new products, then R&D by itself may play the major role in economic success. But, increasingly, this is not the case. Technology diffuses rapidly across international boundaries and technological isolationism is currently in disrepute as a winning business or national strategy.

Thus, the emphasis in commercial R&D must shift to include not only product design but increasingly also the application of science and technology to all industrial operations, including manufacturing, inventory control, quality control, distribution, marketing, and user support and services. The importance of process R&D and an integrated systems engineering approach in both manufacturing and service industries (broadly construed to include transportation and communications) has achieved the status of a great and newly revealed truth in the United States. Many observers have, for example, attributed the U.S. loss of the integrated circuit dynamic read-only memory (DRAM) chip market to Japan mainly to the relative neglect by the U.S. manufacturers of process technology and have argued that, even if all the economic and trade policy factors had been eliminated, the loss would still have occurred. The message has gone out that, in many cases, the way to beat the inventor and innovator of a new product technology is to leapfrog over him with better process technology and product realization. This, of course, is not so much a new discovery as a rediscovery of the process by which the United States achieved many industrial successes utilizing European, and especially United Kingdom, scientific and technological discoveries and developments in the nineteenth century.[8]

The U.S. military played a significant role in the early stages of integrated circuit development as a buyer and supporter of process engineering (including pilot production lines) and developer of applications technology. However, the military influence on the U.S. commercial microcircuit industry has been relatively small for almost two decades (less than 10 percent of sales) and it is difficult to argue that the military impact on the industry has had much to do with its loss of competitive position in commercial sales. On the other hand, military R&D could hardly have done anything directly to affect the competitive position of the industry over the period in question, since process technology for the DRAMs in production was not considered an important or even proper objective of military R&D programs.

As the concepts of product realization have bubbled up in the civilian economy in the U.S., the military weapons acquisition community has recognized that there was much in its outlook, methods, and technical approach that could also benefit the economy and quality of military products, and there has come to be increased emphasis on weapons design, developing, manufacturing, and operation as an integrated system — a natural complement to the already well-developed military integrated logistic systems and to the systems approach to the weapons themselves which has been utilized and improved continuously since World War II.

Military contributions to the development of systems engineering and management (e.g., PERT critical schedule path analysis) to manufacturing process R&D, especially for advanced materials such as titanium and high-strength composites, and to advanced quality control techniques, instruments, and procedures have been considerable, but their transferability to civilian industry apparently has not been large across the board. In some industries, such as aircraft and aircraft engines, the evidence of transfer is large and significant and in others, such as electrical machinery and chemicals, there are few obvious transfers. Generally, the data and methodologies for tracking the diffusion of technologies across technical fields and industry sectors are simply not well enough developed to provide reliable information of the kind needed to address our questions.

Conclusions

The new research on the ingredients for successful commercialization of products resulting from R&D in science and technology have clearly shown that being first with new science and technology, whether it results from military or civilian R&D, does not necessarily lead to achievement of success in the marketplace. Factors having to do with appreciation of customer needs and values in product performance, quality, and reliability and the structuring of production, distribution, and marketing organizations and managements to respond to those factors are often of greater importance, particularly for consumer products. Therefore, it is not possible to judge the so-called spin-off of military R&D to the civilian sector of the economy simply by attempting to assess the success or failure of products for which there is related military R&D in the marketplace.

This view of the workings of the civilian economy is by no means new, but undue emphasis on science and technology in isolation by the U.S. academic, industrial, and political leadership in the years since World War II had, until recently, tended to obscure and downplay the importance of nontechnological factors. However, just as military R&D cannot be blamed for shortcomings of the commercial sector in high-technology products, neither can military R&D by itself be given credit uncritically for the commercial success of high-technology products to which military R&D may have contributed. On balance, military R&D in the technology base programs (for example, those not concerned with engineering development of operational systems such as the B–1, the M–X, and the Trident submarine) seems to have contributed substantially to those fields where the results apply, such as aerospace and computers. This trend has prevailed since World War II, and there is no indication that a change is in process.

In spite of the reasonably good effectiveness of technology interdiffusion and civil technology of DoD R&D in the industrial sectors to which they apply, there is no reason to believe that spin-off from defense R&D can be as effective in meeting civil needs as R&D directed specifically at civil sector requirements. Spin-off should therefore be regarded as a side benefit and not as a reason for doing defense R&D.

The interdiffusion of either civil or military R&D across fields of technology and industrial sectors is poorly understood, although it is clear that such interactions take place in both product and process technologies and that they are of growing importance, especially in the global economy that is evolving. Further research in the interindustry flow of both product and process technology — perhaps in a form vaguely conceived of as a sort of input-output table for R&D products — would help to comprehend the "black box" relationship between technological R&D and the economy that is at the root of many of the questions concerning the impact of military R&D on the civilian sectors of industry.[9]

NOTES

1. Michael E. Davey and Genevieve Knezo, June 15, 1987. Background material for hearings on The Federal Contribution to Basic Research. Prepared for the House of Representatives Committee on Science, Space and Technology, Subcommittee on Science Research and Technology. Congressional Research Service. pp. 24-27.

2. U.S. Department of Defense , Office of the Deputy Undersecretary of Defense for Research and Engineering, Basic Research Programs, 1984. p. 4.

3. Conference Report on "Strategic Defense: Industrial and Political Implications," October 18–19, 1986. Sponsored by the Institute for Foreign Policy Analysis, Inc. Cambridge, MA, and the Institute of Security Studies, Christian Albrecht University, Kiel, Federal Republic of Germany. Conference held in Weikersheim, Baden–Wurttenberg, Federal Republic of Germany.

4. Omnibus Defense Authorization Act, 1984; Committee on Armed Services, U.S. Senate, July 5, 1983, pp. 136, 140, 153.

5. Budget of the United States Government, Fiscal Year 1988, Appendix, Executive Office of the President, Office of Management and Budget, U.S. Government Printing Office. Washington, DC.

6. Committee on Science and Technology, U.S. House of Representatives, Task Force on Science Policy, September 1986. "The Nobel Prize Awards as a Measure of National Strength in Science," *Science Policy Study Background Report* No. 3.

7. "Japanese Technology Today," *Scientific American* (Advertising Section), November 1985, p. J-18.

8. Habakkuk, 1962, *American and British Technology in the Nineteenth Century* (Cambridge University Press, London), pp. 201-202.

9. Nathan Rosenberg, "The Impact of Technological Innovation: A Historical View," in R. Landau and N. Rosenberg, eds., *The Positive Sum Strategy: Harnessing Technology for Economic Growth* (1986: National Academy Press, Washington, DC), p. 20.

5

Improving Weapons Acquisition*

Jacques S. Gansler

The procurement of weapons for our armed forces has traditionally been an issue left to "the experts." However, as the build-up in defense expenditures approaches the trillion-dollar level — the largest in peacetime history — more and more questions have begun to be raised about whether taxpayers are getting their money's worth. Issues range from whether we are buying the right systems — highlighted by the controversy surrounding radios that did not allow the Army to talk to the Navy during the Grenada conflict — to the glaring newspaper "horror stories" about weapons that do not work and grossly overpriced spare parts, coffee pots, and hammers. With the deficit rising dramatically and the perception of chaos and corruption in defense procurement increasing, defense expenditures have begun leveling off,[1] and congressional and executive branch attacks on the defense industry, fueled by the press, have risen in intensity. Meanwhile, efforts have been made by the Pentagon to shift public attention away from government management issues and to questionable actions said to have been taken by suppliers; these have given many the impression that acquisition management has become primarily an issue for auditors and lawyers.

Congress has attempted to address the situation by issuing hundreds of new procurement reforms aimed at correcting the apparent abuses. Many claim that this legislative whirlwind has reached the level of absurdity. For example, under Senate Bill 1958, introduced on December 17, 1985, "no funds appropriated to or for the use of the Department of Defense (DoD) may be obligated or expended for the procurement of any plastic toilet cover shrouds, identified as toilet assembly-

---

*This article is reprinted by permission of the Yale Law & Policy Review where it appeared in Vol. 5:73, 1986.

#941673-101, at a unit cost in excess of $125.00."[2] Additionally, Congress has involved itself more and more in the detailed management of each and every one of the more than 2,000 defense budget line items, annually changing more than half of them in one way or another and requiring for many others that detailed studies be done by the DoD and submitted to Congress. Thus, instead of "getting government off the backs of industry," the trend has been toward increased auditing (by the General Accounting Office, the Inspector General, and others) and greater regulation of defense contractors.

Fortunately, paralleling this mainstream focus on "fraud and abuse," there has been a broader, and far more important, increase in concern about "waste." This is essentially a new look at the effectiveness and efficiency we realize from our defense dollars and at the broad structural changes that are needed to increase the cost-effectiveness of our expenditures. At the beginning of the Reagan Administration, the Defense Department recognized the need for such changes and took the lead in initiating reform. The so-called "Carlucci Initiatives" were a set of acquisition reforms proposed by then-Deputy Secretary (now Secretary of Defense) Frank Carlucci and aimed at correcting many of the historic abuses in the procurement system. They focused on increased long-term efficiency through such changes as greater program stability, improved production management, and establishment of greater realism in the estimation of program costs.[3] While suggesting highly desirable changes, such proposals ran up against the traditional way of doing defense business and were hard to implement, especially in an environment in which everyone was in a hurry to make short-term "fixes."

By the end of President Reagan's first term, however, the movement towards broad structural reform had gained momentum. The need for change permeated all levels of the DoD, including the Office of the Secretary of Defense, the Joint Chiefs of Staff, and the military services, as well as the defense industry. All of the major DoD processes were affected, including the requirements process, for weapons selection and specification; the planning, programming, and budget process, for resource allocation; and the procurement process itself. It is on this broader set of changes that we must now focus.

Background

In early 1985, a bipartisan report on the findings of an independent, eighteen-month study on defense organization was released, emphasizing the need for broad structural change.[4] The members of the study panel included representatives from Capitol Hill, as well as many former Defense Department officials and military leaders. Additionally, the recommendations of this study were endorsed by six former Secretaries of Defense who, in their introduction to the report, stated "[t]here are serious deficiencies in the organization and managerial procedures of the U.S. defense establishment."

This study was followed by the release of a similarly detailed investigation by the Senate Armed Services Committee.[5] This also recommended broad institutional changes; it was supported by a bipartisan coalition led by Senators Goldwater and Nunn. On the House side, Representative Les Aspin (Chairman of the Armed Services Committee) initiated a far-reaching set of hearings on a wide range of defense procurement and management issues. Finally, in February of 1986, the President's Blue Ribbon Commission on Defense Management (the so-called Packard Commission, named after its chairman, industrialist and former Deputy Secretary of Defense David Packard) released its set of preliminary recommendations on defense reorganization and procedural changes.[6]

Again, the focus was on broad, structural changes in the acquisition process itself rather than on the narrower issue of fraud and abuse. The recommendations centered on the need for a new planning and budgeting system, significant reorganization of both the Office of the Secretary of Defense and the Joint Chiefs of Staff, and major changes to be initiated by both Congress and the DoD in the acquisition process, as well as in specific weapons management and buying practices. The Packard Commission continued to issue reports over the next four months (including an "Acquisition Report" in April and a "Final Report" in June),[7] and its recommendations received wide support from both the executive and legislative branches.

Congress and the executive branch are faced with a choice. They can continue their attack on fraud and abuse through greater regulation, stepped-up auditing, and yet more congressional hearings; alternatively, they can shift the debate to the higher plateau of broad, structural reform of the way the Department of Defense will do business over the coming years. Fortunately, it appears as though the emphasis is shifting toward the latter approach, addressing the more significant issues of where billions of dollars can be saved and additional billions redirected to strengthen the nation's defense posture more effectively. Many members of Congress, as well as the president and the secretary of defense, have begun to implement the needed changes. The Goldwater–Nichols Act on Defense Reorganization[8] (which Senator Goldwater said on his retirement in September 1986 was the most significant thing he had done in all his years in Washington) and the President's directive to the DoD to implement the Packard Commission's recommendations — even if it goes against the system — are but two examples of the movement toward change. Perhaps this time it will happen!

Obviously, this article argues for the broader, more significant approach to change. However, the argument comes with a warning: it is possible to go too far, seeking change for its own sake and thereby throwing away the good with the bad. One after another independent studies comparing defense management with the management of other government agencies at federal, state, and local levels have found that the Department of Defense is one of the best managed — if not *the* best managed — of all government agencies.[9] This is easy to see when defense overruns are compared with those occurring in other agencies' major projects. For example, Defense horror stories pale in comparison to cost, schedule, and management problems encountered in the building of mass transit systems, congressional office buildings, superdomes, and so forth. For this reason, while making necessary and dramatic changes in the way Defense conducts business, it is important not to ignore the many important lessons we've learned in buying defense weapon systems over the past forty years. Nonetheless, there is much room for improvement, as the following discussion of current problems makes quite clear.

Problems in the Acquisition Process

There are essentially four main adverse trends which must be reversed if the DoD is to get its money's worth and public confidence in the management of defense spending is to be restored. These are (1) increasing problems with the choice of weapon systems, (2) rapid growth in the cost of defense equipment, (3) lengthening of the acquisition cycle, and, (4) growing problems in the U.S. defense industrial base. Each of these four trends is discussed herein; the subsequent section outlines the changes necessary to counter them.

A. Increasing Concern About the Choice of Weapons Systems

There is a widespread perception that the existing institutional structure of the Department of Defense does not provide for the selection and development of the most cost-effective weapons. This concern is evident, for instance, in the extensive debates over the past several years regarding the $20 billion requested for 100 M–X missiles to put in the old Minuteman silos, a second $20 billion for 100 B–1 bombers to fill the gap until the Stealth bomber enters the inventory, and a third $20 billion for two additional aircraft carrier task forces for possible use against Soviet land forces. The issues are not whether these weapon systems are desired, but rather whether they represent the best way (among many possible alternatives) to spend $60 billion in enhancing the nation's security and, even more importantly, how these spending decisions should be made. Similarly, there has been much debate over, but again, no clear consensus on, the allocation of incremental defense funds. For example, should a 600-ship Navy, more Army units, or more Air Force fighter wings have the highest priority? The difficulty of these questions of strategy — and the resultant weapon selections — is compounded by interservice rivalry for resources. It is further complicated by the fact that technological change offers the opportunity to make revolutionary shifts in the composition of the military forces. Such shifts could multiply relevant military capabilities, if the new technology could be absorbed by the affected military institutions. However, proposals for this kind of dramatic change often fall into ambiguous areas outside of the accepted, traditional service equipment and missions. It is difficult for the armed services to accept such proposals, in a cultural sense, because doing so would mean that long and ingrained traditions would have to be altered or scrapped. For example, the Navy might be able to carry out major portions of its mission of denying use of the surface of the seas to an enemy by using reconnaissance satellites and land-based missiles rather than ships. But such an idea is so foreign to traditional notions of naval operations that it, like similarly novel ideas, receives little attention.[10] Instead, we continue to concentrate on building improved versions of traditional platforms: ships, planes, and tanks. Moreover, the armed services insist that each item of equipment be the "best possible." This insistence is one of the factors leading to the second of the adverse trends.

Rapid Growth in the Cost of Defense Equipment

The U.S. clearly has kept its military equipment at the forefront of the technological state of the art. The cost of this improved performance, however, has been increases of around 6 percent per year in the unit price of each new generation of equipment.[11] The cost of a single ship is currently measured in hundreds of millions and even billions of dollars, an individual plane in the tens or even hundreds of millions, and each new tank in the millions.[12] Under any realistic projection of resources likely to be made available for defense, if unit costs continue to increase, the nation will be able to buy fewer and fewer weapon systems each year.

The armed forces have historically been too optimistic in estimating the cost of weapon systems, especially when first requesting funds for system development. The hope has been either that costs will, in fact, be unexpectedly low, or that more money will become available in the future. More cynically, some suggest that unrealistically low cost estimates reflect a bureaucratic tactic of getting the development program started, but leaving the problem of how to pay for it to those in office in later years. Indeed, as weapons are actually developed and procured, far too often their realized costs have been significantly higher than the initial estimates. The historical average of this program cost growth has been between 50 and 100 percent of the original cost estimate of each weapon system.[13]

Naturally, if there are only a certain number of dollars available for buying a given system and its cost doubles, we can afford to buy only half as many. While the U.S. has been buying more and more capable weapon systems, then, the result of both types of cost growth (from generation to generation and between initial estimates and final price tags) has been the purchase of fewer and fewer systems each year. For example, in the 1950s, we bought around 3,000 fighter planes each year; in the 1960s, the number purchased declined to 1,000 per year; and in the 1970s, we were down to buying only 300 fighter planes per year. Norm Augustine[14] has pointed out that a continuation of this trend would result in our building one fighter plane per year in the year 2054. Even with outstanding individual weapon performance, there is still a minimum quantity of weapon systems which is absolutely critical for the successful completion of any military mission. Maintaining this minimum becomes even more important as the Soviet Union steadily improves the quality of its weapons while still maintaining equipment stocks and production rates that are very high compared to American defense numbers. Thus, these cost-induced reductions in the quantity of U.S. weapon purchases could be devastating. Compounding this problem is the next of the undesirable acquisition trends.

The Lengthening of the Acquisition Cycle

The increasing complexity of modern weapon systems contributes significantly to the lengthening of the time span from initiation of development to completion of production. However, two other, even more important causes of delay are (1) stretchouts of the acquisition cycle resulting from an increasingly burdensome and indecisive managerial and budgeting process in both the executive and legislative branches; and (2) stretchouts resulting from program cost growth and budget reductions. It used to take five to seven years to acquire a weapon system, but new systems now often take twelve or even fifteen years to move from exploratory development to initial deployments in the field.[15] Even after development is complete, the high cost of each weapon dictates that only a few production units can be purchased each year. This means that the deployment of any significant number of a given weapon is still further delayed. The lengthening acquisition cycle thus has a compound effect on the military. First, it results in a decline in America's technological advantage over the Soviets, since most of the systems deployed in the field by the U.S. are older designs. Second, the longer cycle itself reduces efficiency in the acquisition process, which results in still greater unit costs and still lower quantities. Adding to these undesirable weapon system acquisition trends — and to a considerable extent caused by them — is the fourth of the adverse trends.

Increasing Problems in the U.S. Defense Industrial Base

With the long-term decline in rates of production, one would expect to see the industrial base drying up. In fact, during the dramatic shrinkage in defense procurements in the early 1970s,[16] the large prime contractors remained in business by building equipment at very low rates, for instance, one aircraft per month in extreme cases. At the same time, subcontractors and suppliers of parts simply disappeared. Later — during the buildup of the Reagan years — they were often replaced by foreign suppliers, particularly in many critical electronics areas.[17] A series of reports in late 1980 all indicated that significant problems had developed in the U.S. defense industrial base.[18] These studies identified areas of substantial inefficiency for normal operations in peacetime, as well as critical bottlenecks (for instance, in selected essential parts and production equipment). The inefficiency and bottlenecks resulted in an almost total lack of capability to respond rapidly to any emergency condition with a surge in production.[19]

Reversing these four undesirable acquisition trends can be accomplished neither quickly nor easily. The complexity and magnitude of the defense acquisition system does not lend itself to simple solutions. However, partly out of frustration and partly to respond to the public and press clamor for corrective actions, quick fixes have been the approach pursued by both Congress and the DoD. For example, last year Congress considered over 150 different defense procurement reform bills; many of these would, in fact, have been counterproductive and even self-contradictory. Meanwhile the DoD, to correct the "spare parts problem," hired literally thousands of people. The Air Force alone added over 3,000 people just to check spare parts prices. The net result of this "fix" was to double the processing time for ordering spare parts and to actually *decrease* force readiness. Another example of the mismatch in resources is the fact that the DoD now has 30 to 40 percent of the government's plant representatives and auditors checking on spare parts, which actually represent only 3 to 4 percent of the total DoD acquisition dollars. Clearly, attention must be focused on the large-dollar items (the weapon systems) rather than on the small-dollar items such as spare parts costs if significant impacts on defense procurement are to be realized. The small-ticket items, unfortunately, are the ones that appeal to the press, so they have received a disproportionate amount of attention.

Needed Changes

If these significant and undesirable trends in the Department of Defense's acquisition practices are to be reversed, four main changes are required. In order of priority, these are (1) improved long-term resource allocations and weapon system selections (there's no value in properly buying the wrong systems); (2) improved stability in programs and budgets (how can you possibly manage efficiently if the programs, and the dollars allocated for them, are continuously changing?); (3) a shift from the current system that regulates quality and costs to one that creates natural incentives for higher quality and lower costs (it's harder to get people to do things right by directive than if they want to do them by choice); and, (4) greater emphasis on the importance of the health and responsiveness of the defense industrial base (its role as a vital part of our national security must be recognized and steps taken to revitalize the "arsenal of democracy"). These four changes are closely interrelated; all four are required if there is to be significant change in the defense culture and in the way DoD does its business, and if the previously adverse trends are to be reversed to any significant extent over the coming years. To see what is required, we must look at each of the four in greater detail.

Improved Methods of Allocating Defense Resources and Establishing Weapon System Requirements

Traditionally, weapons and other equipment are selected almost entirely separately by each military service, acting independently.[20] The Army, Navy, Air Force, and Marines each choose the systems that appear best suited for their unique, historical missions, according to their own perceptions of requirements. In this way, the armed services design the structures of their forces almost as if they intended to fight independent land, sea, air, and amphibious wars.[21] As President Eisenhower emphasized in 1958 and military experts still agree, future battles will be fought with integrated forces. Clearly, weapons and equipment should be selected to complement one another, so as to maximize the combined capabilities of the armed forces.

Unified military resource and equipment planning would require a strengthened Chairman of the Joint Chiefs and an organization (and staff) more independent of the services, as was suggested (although not implemented) by President Eisenhower in 1958. Long-range, mission-area resource plans would be generated that would make more effective use of the total resources available. Additionally, these integrated plans could take better advantage of the changes in technology that are now available, but that currently result in considerable overlap in the traditional roles and missions of individual services. Such long-range plans, closely tied to military mission future needs, would place an "affordability constraint" on deciding which future weapon systems would be developed and procured, and in what numbers. The Chairman would also recommend a military strategy that would be tied to these resource plans, a link which, many have noted, is currently missing. Such a plan would not only have the military making explicit trade-offs between quantity and quality but would also require explicit trade-offs between dollars for force modernization and dollars for force readiness (trade-offs that are more difficult today, when the Operating Commanders do not have a strong voice in spending decisions but the advocates for new weapons are well represented).

Many of these specific changes are contained in the earlier-noted Goldwater–Nichols Act on Defense Reorganization passed in September 1986. The overall effect of the changes in this Act, when and if it is fully implemented, would be to shift towards *centralized decisionmaking* and *decentralized implementation*. The decisionmaking (on weapons, resources, and strategy) would be done much more on an integrated, multiservice basis, although the individual services would have full authority and responsibility for the execution of the weapons systems developments and procurements, as well as their subsequent support. A long-range, integrated resource plan (generated by the Office of the Secretary of Defense and the Chairman of the Joint Chiefs of Staff) would then form the basis for the second needed defense acquisition reform.

Greater Program and Budget Stability

The United States is one of the few if not the only nation in the world to run its defense establishment on an annual budget cycle.[22] Single-year defense budgets encourage the services, the administration, and Congress to meet annual total budget limits by simply pushing everything out to "next year," that is, stretching out the purchases of most weapons over several additional years — a decision which is far less difficult politically than actually canceling an entire weapon program. Such stretchouts are shortsighted, as they force contractors to produce equipment at inefficient rates of production, resulting in higher unit costs and the ultimate procurement of fewer systems overall. For example, the three-year production stretchout of the F–15 aircraft in the mid-1970s resulted in a two billion dollar increase in program costs (excluding the effects of inflation). Eighty-three fewer fighter aircraft were purchased than would have been possible for the same dollars had the original plan been adhered to.[23]

Thus, Congress could make a very significant contribution to controlling procurement costs by adopting a multiyear defense budget.[24] Two- or three-year budget cycles would introduce the greater stability necessary for contractors to plan more efficient production rates and lower the unit costs of new systems. Furthermore, the stability of multiyear budgeting would encourage the use of multiyear procurement contracts — a far more efficient technique. Finally, and most important, multiyear budgets would encourage the Defense Department and Congress to consider more carefully the long-term fiscal and strategic implications of procurement decisions. Naturally, multiyear budgets could be reviewed at any time in case of changes in political or economic conditions.

A second beneficial congressional reform would be to reduce the number of committees involved in the Defense budget process. In 1983, Defense Department witnesses testified on the 1984 budget before 96 committees and subcommittees; 1,306 witnesses provided 2,160 hours of testimony.[25] These redundant hearings are time-consuming for both Defense management and for Congress and focus extensive attention of the details of the budget rather than on the more important policy issues. Additionally, the many small changes that result from this process — over 1,000 line items are changed each year — introduce great instabilities into defense programs and create added uncertainty about future funding levels and schedules. An Air Force study estimated that savings of 20 percent could be achieved, after a few years' time, by stabilizing the Department of Defense budget.[26] Congressional reforms would help to make these savings possible — but again, such reforms are currently meeting active resistance.

The Defense Department also could help achieve greater budget stability by recognizing how much it hurts itself by not making realistic estimates of total program costs. Historically, the probably significantly higher cost associated with the *risk* of developing new, advanced-technology weapons has not been included in most initial estimates, because potential price tags look far more attractive without these contingency dollars added on.

Inherent in the concept of program stability are two basic requirements, neither of which has been followed very well in the past. First, never start a program until you're ready. That is, don't initiate full-scale development unless the technology has already been demonstrated, firm requirements have been established, the program's likely cost is understood, and relevant operational concepts have been settled. Second, recognize that, historically, a commitment to full-scale development is a commitment to production. No programs have been cancelled by the services after full-scale development has begun, and very few have been cancelled by any other authority, either in the executive or legislative branches. So, subsequent years' dollars must be available for production (an "affordability" test) and all the planning done early on to have a smooth transition from development into production. Proper production planning must be done well in advance to achieve the desired efficiency of production operations. Such planning and commitments would allow both efficient production rates and efficient multiyear contracting.

The first two reforms in defense procurement — improved planning and selection of weapon systems, and stability in programs and budgets — are the highest priority. With these two reforms in place, it would then become possible to manage each weapon procurement program far more efficiently and effectively. However, the additional two reforms to be discussed are also necessary for real efficiency. The combined effect of all four will achieve the needed overall "cultural change" in the way defense carries on its business.

Shifting from Regulations to Incentives in Order to Achieve Higher-Quality and Lower-Cost Equipment

Here we come to the area of specific procurement reforms — the types of measures with which most discussions of "getting more bang for the buck" normally start. In this case, there are two choices. The first choice is for the defense industry to become a totally — and officially — regulated industry. In this case, a plethora of new laws and regulations would be issued covering every detail of defense costs and decisionmaking. The alternative is to create a new environment, one in which the government and its contractors have natural incentives, such as promotions, profits, increased sales, and professional pride, to be efficient in weapons procurement. The result of such incentives would be that government and industry managers would actually *want* to figure out ways to improve the quality and lower the cost of their products. Such market incentives are normal in the commercial world; however, as will be seen, they rarely exist in defense procurement today. They must be created in order to achieve the necessary change. Consider the following ten specific techniques for creating the natural incentives needed for improved efficiency and effectiveness of defense resource management.

Enhanced Professionalism

Experienced government managers are crucial to the successful acquisition of new multibillion dollar, high-technology, high-risk weapon systems. Incentives must therefore be created for DoD to retain, and, even more importantly, to promote effective military and civilian personnel in the agencies that manage the acquisition process. Historically, such stability and rewards have not existed. For the military, promotion potential has lain elsewhere, in operational positions such as squadron commander or ship captain. Rotation rates are high in acquisition management billets,[27] and inexperience, or even no experience, is common. In recent years, however, the Air Force has made some significant strides in the right direction, and in 1985, Navy Secretary Lehman directed that 40 percent of all future admirals must come from the acquisition community. Similar upgrading of status and promotion potential is also required on the civilian side. The first and most obvious step is a reclassification of procurement personnel from "administrative" (essentially clerical) to "professional" categories. This would entail a number of corresponding changes in requirements regarding education, training, and experience. Until the personnel system is fully reformed on both the civilian and military sides, it will be difficult to recruit and retain the most capable and experienced people in the acquisition process. It is unlikely that effective management can be achieved without such changes.

Increased Program Management Authority

Over the years, "cults" of technical specialists (for data requirements, military specifications, schedule control, and so on) have proliferated on the staffs of the services and of the Secretary of Defense and have acquired extensive authority. A program manager is told exactly what he must have in his program by, for example, the "competition advocate," the "reliability expert," the "logistics expert," and the "military component specification expert." Because each of these many individuals has veto power over a program, the program manager must agree to meet their diverse requirements if he wants his program approved — regardless of the costs. In addition, in order to "sell" a program on up the line, the program manager must go through innumerable sets of reviews by management and staff at higher levels. Often over forty sets of briefings will be held for one decision on how the program will be run. A look at either commercial programs or well-run defense programs shows that what successful programs all have in common is a strong program manager — one with full authority to get the job done and with the full support of those senior to him who can force the system to allow it to happen. By contrast, defense management "layering" has built up to such a point that in some cases it is nearly impossible for anything to be accomplished. For this reason, in 1985, Navy Secretary Lehman took the dramatic step of eliminating a whole layer, or organization, by removing the Chief of Naval Material and all of his supporting staff of hundreds of people. This also had the desirable effect of eliminating much of the even larger staff in the next layer down, whose principal job had been supplying data to the eliminated upper layer. Lehman also streamlined the reporting chain from the program manager directly upward. Program managers were thus given both the authority and the responsibility associated with their job and were free to manage in the most effective and efficient fashion practical within the limits set by program dollars. This kind of streamlining is an essential step in more effective acquisition management.

Continual Competition

In the nondefense world, the continual availability of some form of alternative — competition between two or more suppliers for the same product or between at least two different products for the same need — is the normal context of business. Such competition ensures continued, effective incentives to lower costs and improve performance. By contrast, the normal approach for the Department of Defense is to have only an initial ("one-shot") competition for the engineering development of a weapon system. The winner then becomes the sole firm to build that particular equipment for the many years of the program's life. Thus, an F–15 fighter can only be bought from McDonnell–Douglas, and no other aircraft will do the same job as the F–15. The thousands of subsequent technical and program changes required by the government are then bid in a monopoly environment, making even the initial competitive bid on the contract less than useful.[28] All subsequent contracts for the weapon system in question, especially major production contracts, are also bid on a sole-source basis over the next ten to twenty years. The DoD must figure out ways to shift from this sole-source environment to some form or forms of continual, competitive alternative. In the commercial world, if one supplier raises his prices significantly, the buyer simply switches to another; in the DoD world, with only one supplier of a badly needed weapon system, the only option is to buy fewer systems this year and the rest a few years later — at even higher prices.

While continuous competition may not always be practical due to the importance of economies of scale (for example, having two firms building nuclear aircraft carriers might not be cost-effective), most of the time it is. Certainly, the fostering of competition should always be considered and striven for. Where it is not possible or practical at the weapon system level, continual alternatives can be used for critical subsystems. The emphasis here — as with all the required acquisition initiatives — *must* be on incentives to achieve both higher quality and lower costs.

Studies have shown that when the Defense Department has used continual competition in the past (for instance, by splitting the percentage of the annual procurement between two producers, the larger share going to the lowest cost/highest quality producer), overall program cost savings on an average of 25 to 30 percent have been realized, along with significant performance improvements.[29]

Congress has made DoD's position more difficult by wording the laws regarding competition in such a way that competition in defense works for the benefits of the suppliers rather than the buyer. In the commercial world, a business firm generally selects a few, previously qualified sources as bidders, then sticks with the good ones for future business, dropping the ones not providing high-quality products. In contrast, current law requires DoD to accept bids from *all* potential suppliers, regardless of demonstrated quality or prior performance. Congressional regulation of the form of DoD competition goes still further by adding a large number of other factors into DoD's decisions, socioeconomic variables unrelated to cost and quality of product. These include, for example, a small business preference, an unemployed area preference, a minority firm preference — and there are a total of over fifty such requirements. In this way, the DoD, which should be in the position of a very strong buyer (capable of driving quality up and prices down), is hand-tied by its inability to use normal market forces in its behalf. Congress must either do something to help the DoD use its market position to save public resources or acknowledge what it is doing and adjust DoD's budget for the cost of using defense procurement as a means of subsidizing unrelated social goals. The cost of these subsidies has been estimated at between 15 and 30 percent of DoD's spending.[30]

Increased Use of Commercial Systems, Parts, and Specifications

Until recent years, defense technology was generally far ahead of its commercial counterparts so that paying the former's acknowledged higher costs could be justified. However, in many areas (electronics, for example), this is no longer the case. Better and cheaper equipment is often available in the highly competitive and fast-growing commercial marketplace. Nonetheless, the defense establishment (in both government and industry) has clung to its traditions, insisting upon extensive use of special-purpose equipment and of parts built from special military specifications. The overall result is that the DoD pays dearly — often five to ten times as much — for the specialized nature of its parts and equipment, and yet sometimes gets inferior results. For example, in microelectronics, today's commercial equipment not only is built to withstand environments (such as being mounted on automobile engines) that are as harsh as those stipulated by DoD, but this large-volume, field-tested commercial equipment is also higher in quality, lower in cost, and incorporates much more advanced technology than comparable military equipment. One recent study of comparable electronic systems, including computers, radios, sensors, and displays, found the commercial equivalents to be between two and ten times cheaper, up to five times faster to acquire, generally more reliable, and one to three years more advanced in technology — yet capable of performing as well in the same environments.[31]

It is time for Defense to shift to the selection and use of existing commercial systems, parts, and specifications as its first priority, whenever these commercial equivalents will do the military job. These parts have all met the market test for both quality and price. The DoD will, by making this shift, have all the advantages of the continuous competition of the commercial marketplace without having had to create the market itself. This approach has the added benefit of increasing the integration of the military and commercial industrial worlds, introducing not only far more cost sensitivity into military procurements but also providing the potential for a rapid surge in production. Our capabilities for emergency increases in production would be greatly enhanced if existing commercial production lines could be quickly and easily converted to defense production in periods of crisis.

Reward Good Performance with Future Business

Competitive selections of suppliers for future Defense business are based almost completely on promises contained in the proposals and submitted for a particular award. There is very little automatic, institutional consideration of the actual performance (in terms of quality, delivery, and cost) achieved by a particular supplier on prior programs. The opposite approach is taken in the commercial world, where firms are rewarded with new business if their past performance has been good and are closed out of future business if their performance has been poor. Secretary McNamara tried to implement a performance-based, source-selection system when he first came to the Defense Department, but his efforts met with considerable institutional resistance and were subsequently dropped. Nevertheless, the need to reward success has not diminished, and new efforts should be made in this direction.

Profit as an Incentive

Commercial firms are motivated to increase their profit margins by cutting costs. The Defense Department, however, follows the perverse practice of negotiating a contractor's profit margin each year without regard to how the product's costs in prior years have compared to expected costs. In the largely sole-source environment of defense procurement, the contractor thus actually has a built-in, long-term incentive to raise his costs. This follows from the fact that the cost basis used for profit negotiations in production programs is that of the previous year's costs. The higher the costs, the more profit dollars next year, since the percent profit margin tends to remain about the same from year to year. A far better approach would be for the government to allow a higher profit margin in subsequent years if the costs incurred in prior years fell below expectations. If costs rose in one year, the contractor would receive a smaller profit the next; but if costs actually fell in one year, the contractor could be assured that his profit margin would rise in negotiations for next year's contract. Similarly, profit margin could also be tied to a system's demonstrated reliability, in order to create an incentive in this area as well. By rewarding good performance with higher profits, the defense market would again move towards normal commercial practice.

Reward Lower Costs with Increased Quantities

The military services' incentives to achieve lower costs could be greatly enhanced by a policy which permitted the services to buy larger quantities, or improve the performance, of those particular systems for which unit costs fell below expectations. Part of the cost savings would in this way be returned to the services, to be used to acquire greater military capabilities. As things stand now, the relevant program office loses the savings if costs are reduced — another obviously perverse incentive. Instead, saving could be used to improve the performance of systems (for example, through increased reliability testing), to buy more of them, or to pay for needed product modifications. A version of this price elasticity incentive (based on the commercial market characteristic of greater sales at lower costs) was used successfully by former Secretary of Defense James Schlesinger when he offered the Air Force a choice between a larger number of fighter wings if lower-cost F–16 aircraft were purchased or a smaller number of aircraft if it chose to buy the more expensive F–15s (for the same total dollars). The Air Force opted for the lower-performance, lower-cost, greater quantity option, so today there are F–16s in the the Air Force inventory. Unfortunately, the concept of increased quantity for reduced costs has never been institutionalized, so the stress on maximum possible performance from each system has continued.

In order to encourage broad quantity/quality tradeoffs, Congress should move to a "mission area" budget, rather than continuing to use today's "line item" budget. For example, there is currently a separate budget approved for the F–15 and the F–16; if a mission area budget were used, Congress would set the dollar level to be spent on "Tactical Aircraft" and the DoD would make the necessary quantity/quality tradeoffs. In this way, Congress would be forced to address *why* defense money is spent, rather than only deciding *where* defense money is spent. Although this would obviously result in more difficult decisions for Congress, the increased rationality of defense procurement would be highly desirable.

Technology Demonstrations

Faster, more efficient major weapon acquisition programs require that high-risk subsystems (such as radars, engines, and computers) that incorporate next-generation technologies should, whenever possible, be developed independently of the complete weapon system. They should then be fully tested before a commitment is made to include them in the overall weapon system. This practical demonstration of new technology prior to application in a weapon system is the proper use of the "fly-before-buy" concept. It would reduce both the cost risk of major weapon system development programs and the time necessary to complete them. When new subsystem technology has been demonstrated, it can be quickly inserted into the overall program (the common commercial practice of standardized interfaces would have to be used) and brought rapidly into the field. This "modification" approach has already proven to be an extremely efficient way of developing new weapon systems (both in the U.S. and other countries), but the current DoD acquisition and budget processes remain structured primarily around the development of complete new weapon systems.

Design to Cost

In the commercial world, advanced technology is used to simultaneously lower equipment costs and improve the performance of new systems. In the defense world, however, advanced technology is used almost exclusively to maximize performance, while costs continue to increase. A comparison of commercial electronic systems, where performance has been increasing dramatically while costs have been rapidly falling, and military electronic systems, where *both* performance and costs have been rising proportionately, points this difference out very clearly.[32] Additionally, it has been estimated that developing the capacity to get the last few degrees of performance out of a weapon system tends to raise defense system costs by 30 to 50 percent,[33] correspondingly reducing the number of weapons which can be acquired. If unit cost were made an important engineering design criterion along with performance, the DoD could then take advantage of new technologies, both to improve the quality of its equipment and to increase the quantities it is able to purchase. Essentially, the DoD would be trading a very small reduction in an individual system's performance for a large increase in the number of systems acquired.[34]

Fund Nontraditional Concepts

In order to encourage the development of new technology that can be used to improve overall military effectiveness in nontraditional ways, it is necessary for the services and the independent Defense Advanced Research Projects Agency (DARPA) to set aside funds for nontraditional systems and technology. Nontraditional technologies will otherwise remain underfunded, as the institutions that control the research process consider them, for cultural reasons, to be a low priority. (The cultural reasons are, again, that these new technologies would cut across and into historical service roles and missions.) For example, both the Army and Air Force have had trouble funding and utilizing remotely piloted vehicles, even though the Israeli armed forces have clearly shown the military value of such unmanned vehicles in combat. If a special allowance (not charged against the services' budget) were made available for the development and demonstration of prototypes of nontraditional systems, a form of internal competition could be set up between improvements in traditional systems and innovative new ways of accomplishing the same task. The innovative alternatives could then be tested against the traditional approaches. To create the proper incentives, it would be best if these nontraditional approaches were pursued by separate organizations within each service and/or if increased funding were given to DARPA.

In each of the preceding ten areas, the attempt is to substitute natural incentives for regulation; it is important to emphasize that these changes will be strongly resisted by the existing system and will not be easy to implement. Yet, the government has already initiated efforts at implementing some of the ten, and many of the others are very similar to practices that are widely utilized in the commercial marketplace. There is thus a large body of "lessons learned" that could be applied to defense procurement. The combination of these changes clearly would result in a very significant cultural change within the Department of Defense, a shift from heavy dependence on regulation for improved performance and lower cost to the use of natural incentives to meet these same objectives. It is very likely that the use of such incentives will, in fact, prove far more effective than the historic regulatory approach.

The three broad recommendations for change previously described have in common a focus on the buyer, or on the "demand side" — from changes in the requirements and planning process, through revision in the budgeting and program management process, to incentives to create higher performance and lower cost in the execution of weapon system procurements. Stopping after only these changes were made, however, would leave out a major area where effectiveness and efficiency could potentially be much improved, namely the "supply side," or the defense industrial base. This brings us to the fourth and last of the broad reforms required.

Visibility into Industrial Base Problems

Historically, the assumption has been that there is a free market operating in the defense arena, one which adjusts to changing conditions, achieves economic efficiency, and supplies strategic responsiveness to the nation's security needs. Unfortunately, this has not been the case. The principal reason is that the overall defense market consists of one buyer and, in many instances, only one supplier. Under these unique conditions, the Defense Department, as the only buyer, has an obligation to concern itself with the health and responsiveness of the defense industry. In order to do this, DoD needs to have some organization responsible for the industry's health and in a position to take action to ensure it. At times, such an office might encourage the establishment of a second or even a third producer to create competition. At other times, it might encourage the awarding of a contract so as to create greater labor stability for increased efficiency. At still other times, it might investigate the critical lower tiers of the defense industry to ensure that efficiency and responsiveness are attained in the supplying of critical parts. The U.S. is the only nation in the world that does not treat its defense industry as a vital national resource. Today, the Defense Department does not have the organizational structure to perform the functions necessary to maintain the health of the defense industry, at either the prime contractor level or the critical lower-tier levels. Specifically, what is missing is government *visibility* into the conditions of efficiency and effectiveness in the sectors of the industrial base which are critical to national security. This visibility can be provided by gathering data in such areas as the amount of competition in given sectors; labor force stability; bottlenecks; capital investments; foreign dependency; long-term research and development; capacity utilization; surge capability; and integration of civil and military production. When provided with visibility into the health and responsiveness of the industrial base, the DoD can then include such considerations in its major acquisition and budget decisions. For example, DoD could make informed decisions, not possible now, on the best time and location to start up a new production line, or on whether to obtain a second supplier to do research in a critical component area, or on where to target investments so as to make possible the rapid surge of a production line in the event of a crisis. These supply-oriented decisions are not now a part of the

DoD's acquisition process, nor is the data base necessary to provide the needed visibility and oversight available. With the addition of such considerations into procurement decisionmaking, however, far greater efficiency and effectiveness could be achieved in this unique marketplace. Additionally, industrial responsiveness could be made a more significant part of the overall U.S. national security posture.[35]

Conclusion

The last few years have been dramatic ones for "defense reform." Much new legislation was implemented, with a clear and unfortunate trend towards increased and more detailed regulation of all aspects of the defense industry. Additionally, the number of actions taken by the Justice Department on suspensions and debarments of defense contractors went up by a factor of four. The shift towards increased regulation and criminalization has been further heightened by actions within the Department of Defense itself, which moved in the direction of turning over greater acquisition management responsibility and authority to lawyers and auditors as a short-term solution to perceived increases in "waste, fraud, and abuse." At the same time that these negative trends have been developing, however, both the executive and legislative branches have initiated far broader and more long-range efforts at the needed basic structural reforms. Activities are underway which promise greater results in these more significant directions.

Which of these conflicting trends — increased regulation or structural reform — will prevail is not yet clear. It *is* clear, however, that they can't exist together; for example, increased regulation and increased competition are just not compatible. Additionally, *how* the changes are implemented can be just as important as which ones are selected. Many good laws have been killed by poor implementation. Finally, both of these approaches argue for significant change, but change *within* the current overall institutional pattern. There are, of course, two other options. Minor adjustments to the current system could be made (if you believe it's working well, why fix it; just correct the abuses), or, at the other extreme, there could be radical change in service roles and missions, in addition to the formation of a single, civilian buying agency (the current system will never work, so let's scrap it and start over). This article argues for a middle-of-the-road approach. Even the set of recommendations contained herein, however, will still result in rather dramatic changes in organizations and procedures and, ultimately, in a significant "cultural change" in the way the DoD does business.

The implementation of these four changes — improved methods of allocating defense resources and establishing weapon systems requirements ("JCS reform"), greater program and budget stability ("budget reform"), shifting from regulations to incentives for higher-quality and lower-cost equipment ("procurement reform"), and visibility into the industrial base ("defense industry revitalization") — will be difficult and will take significant time, especially given that it takes many years to develop and produce a new weapon system. So that the changes are not totally disruptive, then, they must be implemented on a relatively gradual basis.

Mostly, what will be required is a *desire* for broad, positive change on the part of both the legislative and the executive branches.

Today, many on Capitol Hill and in the Pentagon are attempting to achieve procurement reform in piecemeal fashion, from the creation in the Office of the Secretary of Defense of a new Deputy Assistant Secretary for Spare Parts (a "spare parts czar") to the addition of thousands of new auditors, and from hundreds of pieces of detailed procurement reform regulations and legislation to even greater congressional micromanagement of every defense budget line item. Such a chaotic climate makes it difficult for those in Congress and in the executive branch who are trying to make the needed, broader changes take hold. However, the coming years will represent an even more significant challenge for those in charge of defense procurement. No longer can one expect that large increases in defense budgets are likely to continue. In the presence of significant budget constraints, there will be even fiercer political infighting (again, both on Capitol Hill and within the services). In this environment, there are likely to be increasing cries for "immediate reform," with its associated flood of new legislation and newspaper headlines. The challenge will be to effect the required broad changes in the presence of this overall environment, without damaging what is worth saving in the existing system. It is clearly a difficult challenge to meet. If the necessary changes can be made, however, the public's confidence in DoD management can be restored, the taxpayers can get their money's worth, and our national security can be strengthened. These results are worth the extra effort.

NOTES

1. In the first term of the Reagan Administration, defense spending authority, minus inflation, rose 36%. In fiscal year 1986, real defense spending fell by 6%. "Is the Bull Market Over?" *U.S. News & World Report*, Oct. 27, 1985, p. 55.

2. S. 1958, 99th Cong., 1st Sess., 131 *Cong. Rec.* S17765-66 (daily edition Dec. 17, 1985).

3. The Carlucci Initiatives were first issued in April 1981. They were summarized in *Streamlining the Acquisition Process*, Aviation Week & Space Tech., May 4, 1981 and formally published in *U.S. Dept of Defense, U.S. Dept of Defense Directive 5000.1* (Mar. 19, 1982).

4. Center for Strategic and International Studies Defense Organization Project, Georgetown University, *Toward a More Effective Defense: The Final Report of the CSIS Defense Organization Project* (1985).

5. Staff of the Senate Committee on Armed Services, 99th Cong., 1st Sess., *Staff Report on Defense Organization: The Need for Change* (Committee Print 1985).

6. President's Blue Ribbon Commission on Defense Management., Interim Report to the President (Feb. 1986) [hereinafter, *Packard Interim Report*].

7. President's Blue Ribbon Comm'n on Defense Mgmt., *A Formula for Action: Report to The President on Defense Acquisition* (Apr. 1986) [hereinafter *Packard April Report*]; President's Blue Ribbon Comm'n on Defense Mgmt., *A Quest for Excellence: Final Report to The President* (June 1986) [hereinafter *Packard Final Report*].

8. Goldwater–Nichols Department of Defense Reorganization Act of 1986, Pub. L. No. 99-433, 100 Stat. 992.

9. See *Packard April Report, supra* note 7, app. A, pp. 37-38 (summarizing studies).

10. For a more extensive discussion of the emphasis on and resistance to technology in traditional and nontraditional weapon systems, see J.S. Gansler, *The U.S. Technology Base: Problems and Prospects, in Technology, Strategy and National Security*, p. 105 (F. Margiotta and R. Sanders, eds., 1985). For an excellent historical perspective, see E. Morrison, *Men, Machines, and Modern Times* (1966).

11. An increase of 6% per year is arrived at even after adjusting for inflation, as well as for the higher unit price associated with the reduced quantities typically purchased today. See Defense Systems Acquisition Review Counsel Working Group, *Final Report: Weapon Systems Costs* (1972); see also J.S. Gansler, *The Defense Industry*, p. 16 (Figure 12) (1980).

12. The Navy recently estimated the cost of one aircraft carrier at $3.4 billion. *Washington Post*, Oct. 26, 1986, A21, col. 5. An F–15 fighter costs approximately $38 million and a B–1B bomber well over $200 million, in 1985 dollars. (The Air Force has committed approximately $20.6 billion for 100 B–1B bombers, in 1981 dollars.) Each M–1 tank, for instance, costs around $2.4 million.

13. See U.S. Air Force Systems Command, Andrews Air Force Base, MD, Dept. of the Air Force, Dept. of Defense, *Affordable Acquisition Approach* (Feb. 9, 1983) [hereinafter *Affordable Acquisition Approach*]; House Committee on Government Operations, Inaccuracy of Dept. of Defense Weapons Acquisition Cost Estimates, H.R. Rep. No. 656, 96th Cong., 1st Sess. (1979); U.S. Comptroller General, General Accounting Office, Acquisition of Major Weapon Systems; Dept. of Defense (Mar. 18, 1971) (No. B-163058).

14. Norman Augustine, President of Martin Marietta and former Undersecretary of the Army, is the author of *Augustine's Laws*, first published by the American Institute of Aeronautics and Astronautics, Inc. 1982. A newer, expanded version was published in 1986 by Penguin Books.

15. D. Lockwood, *Cost Overruns in Major Weapons Systems* (Congressional Research Service Report No. 83-194F), Oct. 15, 1983.

16. From 1969 to 1975, the annual procurement account dropped from $44 billion to $17 billion (excluding inflation effects). See J.S. Gansler, *supra* note 11, p. 12, fig. 1.1.

17. For a discussion of the growing foreign dependence issue, see J.S. Gansler, "U.S. Dependence on Foreign Military Parts: Should We Be Concerned?" *Issues in Science & Technology*, Summer 1986, p. 17.

18. Defense Industrial Base Panel, House Armed Services Committee, *The Ailing Defense Industrial Base: Unready for Crisis*, H.R. Doc. No. 29, 96th Cong., 2d Sess. (1980); Defense Science Board Task Force, Dept. of Defense, Report on Industrial Responsiveness (Nov. 21, 1980); U.S. Air Force, Dept. of Defense, Statement on Defense Industrial Base Issues (Nov. 13, 1980); see also J.S. Gansler, *supra* note 11.

19. For example, it was reported that it would take over three years for an existing aircraft production line to increase its output significantly.

20. *Packard Final Report, supra* note 7, p. 43. The independent service resource planning and weapons requirements process is discussed on pp. 44–48.

21. This explains why, as noted earlier, in the Grenada incident in 1983, the radios of the Army and Navy operated differently, prohibiting the needed direct communication among them during the conflict.

22. For a discussion of the defense planning process in many other countries, see *Reorganizing America's Defense: Leadership in War and Peace* (R. Art, V. Davis, & S. Huntington, eds., 1985); see also J.S. Gansler, *supra* note 11, pp. 244–256.

23. A detailed discussion of schedule stretchouts, their effects, possible corrective actions, and the specific F–15 example is contained in *Affordable Acquisition Approach, supra* note 13.

24. For a discussion of multiyear budgeting, see J.S. Gansler, "Reforming the Defense Budget Process," *The Public Interest*, Spring 1984, p. 62.

25. Oral Presentation by Vincent Puritano (DoD Comptroller), University of Kentucky (Dec. 2, 1983).

26. For a discussion of the effects of instability, see *Affordable Acquisition Approach, supra* note 13.

27. D. Lockwood, *supra* note 15, pp. 23–24.

28. For example, on the F–111 aircraft program there were a total of about 394,000 changes.

29. For a brief summary of examples of program cost savings, see J.S. Gansler, "There's Precedent for Pentagon Thrift," *L.A. Times*, Oct. 2, 1983, Section 4, p. 5. For a more detailed discussion, see J.S. Gansler & L. Kratz, *Effective Competition During Weapon System Acquisition* (monograph published by The National Contract Management Association) (1985).

30. Oral presentation by Major Gen. Bernard Weiss, USAF, Defense Science Board 1986 Summer Study on Commercial Components (July 20 to Aug. 1, 1986).

31. Defense Science Board Summer Study Task Force, U.S. Dept. of Defense, Report on Use of Commercial Components in Military Equipment (1986).

32. For a discussion of the cost trends (vs. performance and time) of military electronics systems, see J.S. Gansler, *supra* note 11, p. 100, fig. 4.1.

33. For a discussion of the use of technology to reduce costs, see *Packard Interim Report, supra* note 6, pp. 18–20; see also J.S. Gansler, *supra* note 11, p. 279.

34. The Air Force is adopting this technique to achieve lower production costs on its next-generation fighter program (called the Advanced Tactical Fighter). It is to be hoped that this will allow for future purchases of the needed quantities of this aircraft, and will reverse the earlier-noted trend of shrinking annual purchases.

35. A more detailed discussion of the needed industrial base actions may be found in J.S. Gansler, "Can the Defense Industry Respond to the Reagan Initiatives?" *Int'l Security*, Spring 1982, p. 102.

6

Trends in Conventional Arms Technology: Revolutions in the Making?

Alex Gliksman

In an era where change is a constant, the appellation "revolutionary" has become so hackneyed a descriptor of new developments that its currency has been much devalued. Those with a particular fascination for new developments in science, politics, or military affairs — and few of us are exempt from this category — are especially guilty of this tendency toward overstatement.

Fortunately, as compensation, one also finds an equal if not more powerful common propensity to receive such claims with a healthy measure of skepticism. Though the act of discounting is typically done in private, there have been many occasions, especially if issues of great moment are involved, when doubts become the catalyst for important national debates. More often than not, the evidence behind the claims is found wanting.

What follows will look at two major developments in the military affairs field that are having a substantial impact on defense priorities, and may prove to be revolutionary in their consequences: advances in the application of information technologies to military systems and the global proliferation of arms production capabilities. Before examining these developments, the Strategic Defense Initiative (SDI) will be discussed as an example of a "world-altering" new technology whose progress has been less than its revolutionary claims.

"Star Wars" — Lessons from a Revolution That Wasn't

A vision was offered by President Ronald Reagan in his now-famous March 1983 "Star Wars" speech, where he hoped to make nuclear-tipped ballistic weapons "impotent and obsolete." Skepticism greeted that address, as well as strongly worded critiques that have persistently shadowed every Department of Defense (DoD) plan or pronouncement on the resulting SDI. The vision and the skepticism amply illustrate both the impact that promises of radical change can have on national policy[1] and the doubts that often confront such claims.

SDI depicts another aspect of this phenomenon as found among a major segment of American opinion. Americans, particularly those on the political right, have a strong attraction to technical remedies in matters of national security and the methods used to provide it.

Two considerations lie behind this proclivity in American conservative strategic thinking. First, though many on the right have few reservations about confronting international security problems, they prefer that Washington exercises options that are quick and decisive. This makes conservatives ill at ease with diplomacy, a process they find slow to work with results often vague and inconclusive. Second, many of the same individuals also take a narrow view of the ingredients of American security. Emphasizing the word *national* in the concept *national security*, they feel that as a patriotic act, security rests exclusively on a nation's own efforts.

Already disinclined towards diplomacy, those on the right are doubly troubled by that branch of art called arms control. Since arms control depends, in part, on others' good behavior, they find the process illogical at best. But what makes it complete folly to them is the reliance of arms control on the good conduct of not just any other nation, but of a potential adversary.

While this leads some conservatives to yearn for isolationism, it leads many others to want to acquire the military means that would place America's global destiny exclusively in American hands. Given the dramatic effects that technology has had on America's economic and military fortunes during this century, many on the right believe that the United States should again look to its technological prowess to break free of existing security binds. For instance, they see the decisive role played by the atomic bomb in ending World War II in the Pacific, and ask why technology cannot give this nation a new trump card that will minimize, if not end, America's need for international compromise and for messy foreign entanglements.

Thus, when identifying the forces that propelled SDI, one finds that although the prospect of technological breakthroughs was a factor behind the enthusiasm of program supporters, frustration with superpower relations seemingly played a far greater role. Nothing captures this perspective better than President Reagan's words at a 1986 political rally:

> No responsible president should rely solely on a piece of paper for his nation's safety. We can either bet on American technology to keep us safe, or on Soviet promises. I'll bet on American technology any day.[2]

For many SDI supporters, arms control was a failed and potentially fatal experiment. Rather than making a dent in Moscow's nuclear arsenals, the agreements of the 1970s tied American hands, while leaving a deceitful Soviet leadership unfettered in pursuing a weapons build-up. SDI seemed a better alternative. Even if it failed to attain its full potential, the program offered action instead of more talk in response to Moscow's growing arsenal. And, if by chance SDI could achieve its goals, the initiative would provide the means for dealing with the Soviet nuclear threat without having to make deals with the Soviets.

But, despite proponents' best wishes and efforts, five-plus years of relentless scientific, budgetary, and strategic scrutiny have taken their toll. The weight of political opinion seems to have concluded that the objectives of Star Wars are out of reach. In short, there seems little reason to hope that a technological revolution in strategic defense will materialize now or in the foreseeable future.

The Information Age — Implications for Warfare

But just as SDI's fortunes are in descent, claims of yet another revolution in military technology have come forward. Instead of strategic defense, this time the focus is on conventional warfare.

This revolution is said to be already in the making. Its roots are traceable to the first guided bombs used by the United States in Vietnam during the 1960s. Its success rests upon advances in microelectronics, computers, sensors, and software. The most recent developments are transforming what were previously "dumb" military systems — dependent upon the limitations of human senses and speed of calculation — into so-called "smart" and even "brilliant" weapons capable of finding and accurately attacking targets, at distances and with speeds previously unattainable. If research into the more exotic concepts yields positive results, conventional arms could assume missions currently reserved for nuclear weapons. Should this occur, the most basic military distinction, that between nuclear and conventional weaponry, would have to be abandoned.

This technological push has led civilian and military defense decisionmakers to rethink strategy and revise defense research and development (R&D) and acquisition plans. A major focus lies with strategies, concepts, and capabilities for launching conventional "deep strikes" into an opponent's territory. This has begun to erode the boundaries that separate front and rear battlelines. Potentially, every militarily significant point in an entire region could simultaneously come under fire.

It is worth elaborating on several aspects of emerging technology that lead some to believe that it will alter the military environment in a revolutionary way. In the past, the advance of military technology was measured in terms of increased firepower. The advent of nuclear weapons brought this progression to an end.[3] Current military R&D seeks to take warfare down a completely different path. Its aim is to replace firepower with brainpower as the key determinant of the military balance. The key characteristics of emerging technology systems are their ability to collect and analyze data on military activities as they happen — at any one place or worldwide — and to swiftly bring weapons to bear against hostile forces in an extremely rapid and accurate way. If enemy forces can be rapidly identified and attacked with great accuracy, explosive power ceases to be important. Moreover, reducing collateral damage would create a myriad of new military options.[4]

The question arises that if such dramatic and potentially revolutionary changes in the military environment are afoot, why have these developments failed to receive wide attention? The answer is that this is only partially the case. As noted previously, and as will be detailed later in this article, national decisionmakers are acting on these issues. It is the larger, defense-interested public that has failed to take notice. A number of characteristics of military technology in general and these developments in particular, may help explain this.

First, as noted previously, both the public and expert communities have been saturated with claims that this or that development would produce revolutionary changes in military affairs. These claims have proven to be either exaggerations or just plain false. Defense industries eager to sell their products are as much to blame for this condition as anyone. It is the story of the boy who cried wolf. After so many false alarms, few people are willing to listen.

Second, conventional arms have never been the center of attention in the nuclear age. This may be changing as a consequence of the Intermediate-Range Nuclear Forces (INF) Treaty.[5]

Third, the fact that the latest advances deal with information technologies takes the glamour out of some of the most dramatic developments. For instance, many of the changes wrought by the new technology are in the fields of data collection and its management. Capabilities that permit improved coordination of military forces and allow commanders to obtain better performance from existing military assets are an important aspect of this new technology. It is hard to visualize dramatic changes when the impact of these capabilities is measured in terms such as operational effectiveness rather than the more traditional, easily recognized indicators of technological achievement. Thus, the new technology may not make weapons systems travel faster, fly higher, or maneuver more quickly; nevertheless, it will increase the probability that a weapon will find and destroy its target.

Fourth, the application of emerging technologies often involves numerous improvements in capabilities that span the range of military assets and that individually are incremental in nature. They appear dramatic only when examined in total. Unlike the detonation of an atomic bomb, one cannot point to a single event or device that captures the potentially revolutionary character of these developments.

Finally, since a large part of the revolution rests on sensors, microelectronics, computers and software, some of the most spectacular developments are internal to the workings of weaponry and are thus not visible to the naked eye. Viewed externally, a high-tech weapon will look no different than its low-tech counterpart. Nevertheless, the electronic "eyes" and "ears" that can identify tanks and track their movement will be no less dramatic in their effect than electro-magnetic guns capable of hurling artillery shells at five miles per second at tanks.

Applying the New Technology

To repeat, what is involved here is the across-the-broad application of technology emerging in the information revolution to virtually every type of military systems and subsystem. Many of these changes entail improving the performance of existing types of weaponry and/or allowing old weapons concepts to perform new missions.

Examples of systems whose performance has already been, or in the foreseeable future may dramatically be, improved include smart anti-aircraft missiles and anti-tank submunitions. Remotely piloted vehicles (RPVs) are an example of a military system that will benefit from the application of advanced technology. Incorporating new technologies into RPVs will improve the performance of existing missions and allow them to adopt new military roles.

The Stinger shoulder-fired, anti-aircraft missile illustrates the potential effect that the information revolution can have on warfare. By employing sophisticated microelectronics, including an infrared homing system, to do the most exacting tasks involved in attacking hostile aircraft, Stinger leaves the operator with a few relatively simple tasks. The operator first points the weapon at the aircraft, then Stinger takes over, determining whether it is a friend or a foe and telling the operator when to fire. By employing internal complexity to achieve operational simplicity, Stinger may well represent the best in smart munitions.

Its devastating effects on Soviet aviation in Afghanistan probably played a decisive role in the Kremlin's decision to withdraw. Some reports suggest that 50 percent of the Stingers fired by the Afghan Mujaheddin found their mark. Its $85,000-per-round price tag compares favorably with the multimillion dollar cost of each Soviet aircraft destroyed and suggests that high tech need not always mean high price. Thus, without achieving the perfection of "one shot, one kill" that some hope will be possible in the future, Stinger seems very much the ideal weapon, offering a model for the future application of military high tech.[6] Looking toward the future, optimists believe it possible to achieve comparable results with conventional munitions launched hundreds or even thousands of miles from their targets. Smart submunitions are another possibility. Suppose a formation of tanks is spotted — instead of launching numerous strikes against this group of targets, it may be possible to launch a single aircraft or artillery round that would dispense a cluster bomb in the vicinity of the formation. Packed with tens or hundreds of bomblets, individually equipped with miniature sensors and microprocessors, each bomblet in the cluster (such as is proposed for the U.S. Army's Terminally Guided Submunition) would be independently guided toward a tank or, even better, towards a tank's most vulnerable areas, like the engine compartment (as defense technologists hope to achieve in the air-launched, heat-seeking, SKEET submunition) or like the turret (a capability to be provided by the twin Infrared and radar sensors of the Search and Destroy Armor submunition). Instead of one shot, one kill, these submunitions would make for one shot, many kills.

RPVs have long been part of many nations' military arsenals. Some of the best are produced by other nations, notably Israel, who used them with great effect in 1982 to defeat Syrian air defenses. RPVs have the advantages of avoiding unnecessary risks to pilots and of freeing manned aircraft for missions where a pilot remains essential. The new technologies increase prospects for producing RPVs that are smaller, cheaper, and more capable. If these parameters are met, it may be possible to field RPVs in quantities sufficient to give local field commanders independent intelligence and attack capabilities.

The U.S. Army's recent experiment to utilize new technologies, the RPV is a lesson in the risks in grasping for too much high tech too quickly. By using the capabilities inherent in advanced sensors and microprocessors, RPVs have the potential to perform a variety of functions, including surveillance, reconnaissance, target designation, electronic jamming, and attack. But, rather than attempting to develop specialized RPVs for each role, the Army looked to what became known as the Aquila RPV to do much of it alone. The result was a system with dubious capabilities to perform any of its missions, produced at exorbitant cost. A comment Senator William V. Roth, Jr. (R–DE) made to a reporter is worth repeating here: "The only drone we've produced is that of the Army asking for more money."[7] The Army is starting over again.

There is great potential in the application of new technologies to RPVs, but a measured high-tech approach is most likely to obtain results. A U.S. Navy RPV program and an Air Force program to develop an electro-optical sensor that can be deployed on this RPV and on manned aircraft will provide both services with unmanned, real-time reconnaissance capabilities. By exploiting advances in technology that can be applied to a single mission, these programs should lead to fielded systems by the mid-1990s.

Technological advances that will allow familiar forms of military systems to perform new and potentially dramatic missions are exemplified by developments in the field of airborne radar. Existing airborne radar systems have air defense as their primary mission. Incorporating advances in sensors, computers, and software, it will be possible to field aircraft that execute the much more demanding role of acquiring, tracking and targeting land forces against a background of ground clutter. Ideally, such a system would have the capacity to "see" all tank formations operating for hundreds of square miles, assess the course of the battle, determine which tank formations are most threatening, and decide which weapons under its control can best attack them. The Joint Surveillance Target Acquisition Radar System (JSTARS) is slated to have these capabilities. If several outstanding issues can be resolved, notably reducing the vulnerability of this high-value target to attack, JSTARS could enter service in the early to mid-1990s.[8]

The Proliferation of Arms Producers

Although little noticed, the spread of technical know-how has combined with a desire for security autonomy to fuel the growth of advanced, indigenous arms production capabilities throughout the Third World. This is rapidly multiplying the number of players who can affect regional and global security equations and potentially draw in the superpowers triggering global war.

Admittedly, if viewed in terms of the weapons produced, the acquisition of arms manufacturing capabilities by nations in the developing world may not appear impressive. A decade from now, however, Third World nations will have the capability to build arms which are currently commonplace in the advanced industrialized states. But such an assessment belies the significance of the proliferation of arms production capabilities. There is more than enough destructive potential in today's conventional arms. The growth of new production centers would represent a significant departure from the existing international security environment.

With such capabilities in hand, some developing nations may be less hesitant in resorting to force to advance their interests and resolve disputes. The more nations having the capacity to produce weapons, the greater is the potential for deadly regional conflicts. At a minimum, the growth of modern weapons arsenals in these nations will increase the insecurity of neighboring states. Moreover, the United States, the Soviet Union, and other advanced industrialized states will no longer be able to control the movement of arms or settle disputes among themselves. If the advanced industrial states enter the fray directly, their forces would face new and unprecedented dangers.

Current events highlight the risks of ignoring the reality brought on by arms production spread. For instance, Washington's belated recognition that some relatively advanced technologies are no longer exclusively controlled by a select group of states almost rendered moot American efforts to curtail the transfer of ballistic missiles. In the wrong hands, ballistic missiles could be used to deliver nuclear and chemical weaponry. Even conventionally armed missiles would wreak havoc. To promote this initiative, Washington initially focused its diplomacy on the advanced industrialized states. But in doing so, Washington overlooked several developing states, such as the Peoples' Republic of China (PRC), India, Brazil, South Korea, and Taiwan, who are also capable of manufacturing ballistic missiles. The United States has now sought to engage these new producers in the process. The Soviets, who see it as in their interest to keep this technology out of as many hands as possible, have privately endorsed the American plan.

In the future, there will certainly be a growing need to involve the new arms producers in arms control, particularly when indications are that some are motivated by a single-minded quest for export profits without regard for the security implications of these sales. One ominous example is China's arms export behavior. In its search for hard currency to finance modernization, the PRC appears willing to provide arms to all comers, including those in conflicts. The PRC was Iran's supplier of the Silkworm cruise missile and is now supplying ballistic missiles to Saudi Arabia capable of striking Tel Aviv.

Even an end to fighting in the Persian Gulf may, in the long run, prove to be a mixed blessing. While closing this lucrative arms export market may discourage some nations from becoming arms producers, the cease-fire has freed at least one of the belligerents to expand its weapons-making base. Released from the war's demands on resources, Iraq now hopes to establish, in cooperation with Egypt, a modern arms industry on a par with that of Israel. A top priority is the production of surface-to-surface missiles.[9] In combination with Iraq's chemical weapons, this is a deadly formula that could precipitate a sharp response from Israel and encourage Iran to undertake a comparable production effort.

The Politics of a New Military Revolution

Having heard so many false alarms of military revolutions, many readers will likely remain skeptical that these technological changes are truly revolutionary. There are many experienced analysts who share their doubts. These analysts say that systems like Stinger are an anomaly. The norm is that high tech means expensive, and they doubt that the United States can afford to produce advanced weaponry in militarily significant quantities. Similarly, too much of advanced technology weaponry suffers from complexity. Such weaponry is of no virtue when most troops lack the education required to maintain and operate it. Finally, analysts remind us that the military balance is a two-sided game. The Soviets and other potential adversaries will do what is necessary to field effective countermeasures. In the end, the United States will find itself financially spent and as, if not more, vulnerable than before.[10]

These are legitimate issues that must be satisfactorily addressed as part of the R&D effort. As we have often seen, military system advocates will speak of a system's attributes and ignore its problems, as if such shortcomings would resolve themselves. Nevertheless, while acknowledging the importance of these reservations, it is important to recognize the power that perceptions have in international security. Here, we find that, independent of the technical facts, political and military realities are being shaped by the conviction that a revolution is underway. This is true for both Washington and Moscow.

In Washington, the latest promises of a new age in security come with a complete array of security concepts and programs.[11] And, just like SDI, there is a fresh set of buzz words and acronyms to match. Today's concepts are called "competitive strategies" and "discriminate deterrence."[12] The word "initiative," brought into vogue by the Reagan strategic defense plan remains in fashion. Instead of SDI, there are programs called BTI for Balanced Technology Initiative and CDI for Conventional Defense Initiative. Each program seeks to foster research and development of advanced and often exotic conventional weapons concepts.

On the political level, there is a big difference between these efforts and Star Wars. SDI was thrust upon a reluctant Congress by an eager administration. In contrast, many in Congress share the administration's optimism for the promise of high-tech conventional arms. Both hope that the new high tech can provide a qualitative answer to Moscow's quantitative conventional edge. If these new technologies have generated a contest between the legislature and the executive branches, it is over which branch of government can better champion conventional R&D efforts. Thus, while the competitive strategies concept emerged from DoD and discriminate deterrence is the product of an Executive branch-mandated commission, BTI was launched in the Senate and CDI has its origins in the House of Representatives.

In this presidential election year, the views of the Democratic and Republican contenders are important. They will likely determine the priority given these new programs in coming years. And, as is true elsewhere, the Democratic and Republican presidential candidates seem to hold favorable views of these programs. If there is a difference between George Bush and Michael Dukakis, it is on the emphasis the candidates place on these programs. While Bush has endorsed a continuation of Reagan's high tech defense policies, both nuclear and conventional, Dukakis' defense pronouncements have been explicit in supporting high technology for conventional systems. Dukakis has staked a great deal on this issue by promising that, if elected, research, development, and deployment of advanced conventional weaponry will be a centerpiece of his administration's defense program. Thus, independent of who occupies the White House, a focus on the "revolution" in conventional high-tech weaponry will, at the very least, be an important, if not the top, American defense priority.

The Impact in Moscow[13]

Perhaps because of their technological backwardness, the Soviets were quick to appreciate that technology emerging from Western research centers represents a break with the past. Though dismissing SDI's goals as "fantasy," the Soviet have been more sober in their assessment of advanced conventional weaponry.

It was a Soviet Marshal, not a western defense official, who first termed these developments in conventional arms technology a "revolution in military affairs." It is also the Soviets who first voiced the notion that nonnuclear advanced technology weaponry might be "conventional" in name only. General Secretary Mikhail Gorbachev, together with the former and current Chiefs of the Soviet General Staff, Marshals Nikolai Ogarkov and Sergei Akhromeyev, have each expressed this concern with regard to the conventional weaponry they see being developed in the West. Soviet unease is amplified by the inadequacies of the Soviet technological base in responding to this challenge. The Soviets traditionally have used their strength in mass production to redress qualitative deficiencies. But the unique character of emerging Western technology raises doubts in the Kremlin about the continued viability of this quantitative arms approach. The result is a deep sense of military vulnerability. Much of the urgency evident in Gorbachev's plan to radically "restructure" the Soviet economy, including rebuilding its techno-industrial base, derives from this concern.

The Soviets' international initiatives also mirror Moscow's discomfort. Thus, just as rebuilding the Soviet techno-industrial base is the central theme of Gorbachev's domestic policies, slowing or stopping Western arms technologies programs is a key aspect in Soviet arms control proposals.

Some examples: In January 1986, General Secretary Gorbachev announced an arms control initiative calling for the "complete and total elimination of nuclear *and* other means of mass destruction" [emphasis added]. A key, but largely overlooked provision, of this Gorbachev plan would ban R&D on exotic conventional weapons.

As for SDI, it is a concern for research spinoffs rather than strategic defense that is the focus of Soviet proposals. The post summit news conference Gorbachev held at Reykjavik in 1986 provided clear evidence of this. While insisting that SDI defenses did not "bother" him, Gorbachev conceded that "there is a military aspect after all. The SDI can lead to new types of weapons." The Soviet General Secretary and other Soviet officials have repeatedly emphasized this theme in their SDI discussions, and high-tech conventional weapons programs have figured heavily in other arms control negotiations.

Indications are that, in parallel with American efforts to turn emerging arms technologies into new military muscle, Moscow will want to make research, development, and deployment of advanced-technology weaponry the hot issue in U.S.–Soviet security discussion, including the forthcoming post-INF European arms talks.

Footnotes

1. Ronald Reagan's readiness to undertake the most costly defense research program in American history is powerful testimony to the belief in the prospects technological revolution in warfare can have on national policy and international relations.

2. Philip Geyelin, "The Selling of SDI," *Washington Post*, November 9, 1986.

3. It is the almost limitless capacity for devastation of nuclear weapons that makes solving the population defense problem appear so remote.

4. The author recognizes that this also creates serious arms control problems. Given space limitations, these will not be addressed here.

5. See "Trading Nukes for Conventional Weapons: Prospective Euromissile Treaty Would Open New Opportunities," *Christian Science Monitor*, September 3, 1987; and "NATO After INF: Toward A New Alliance Consensus," *National Defense*, December, 1987.

6. The term "one shot, one kill" appears in the title of a *U.S. News and World Report* March 16, 1987, cover story on emerging arms technology.

7. For this quote and a colorful rendition of the Aquila story, see Mark Thompson, "Army Drone Plagued by Problems," *Philadelphia Inquirer*, January 13, 1988.

8. For the most thorough examination of the attributes and limitations of conventional military systems designed to make near-term use of emerging technologies, see Of-

fice of Technology Assessment, U.S. Congress, *New Technologies for NATO: Implementing Follow-on Forces Attack* (Washington, DC: U.S. Government Printing Office, June 1987).

9. Patrick E. Tyler,"Iraq Is Said to Project Broad Role in Mideast," *Washington Post*, August 22, 1988.

10. The thinking of a key critic is presented in Steven L. Canby, "Conventional Weapons Technologies," *SIPRI Yearbook 1987, World Armaments and Disarmament* (Oxford, England and New York: Oxford University Press, 1987), and "New Conventional Force Technology and the NATO–Warsaw Pact Balance: Part I," Adelphi Paper 198, *New Technology and Security Policy: Part II* (London: International Institute for Strategic Studies, 1985). For a detailed technical reviews of the problems, see Office of Technology Assessment, *op. cit.*

11. This section will not address the politics of arms production proliferation. Thorough readers should note that the last two reports cited in footnote 10, provide a sense of Washington's perceptions of this issues.

12. For a recent representation of DoD's competitive strategies concept, see Frank C. Carlucci, *Annual Report to Congress, Fiscal Year 1989* (Washington, DC: U.S. Government Printing Office, 1988), pp. 115-118. The title of the report by the Commission on Integrated Long-Term Strategy, *Discriminate Deterrence* (Washington, DC: U.S. Government Printing Office, 1988) is the source of this expression. Both the Commission's report and that by President Ronald Reagan, *National Security Strategy of the United States* (Washington, DC: U.S. Government Printing Office, 1988) highlight the revolutionary implications of emerging conventional weapons technology and of arms production spread.

13. For more on Soviet concerns, see the author's "Behind Moscow's Fear of 'Star Wars'," *New York Times*, February 13, 1986; also as "Moscow's Real Fear on 'Star Wars'," *International Herald Tribune*, February 14, 1986; "Deterrence Without Nukes," *New York Times*, May 18, 1987; also in *International Herald Tribune*, May 19, 1987; and "Trading Nukes for Conventional Weapons: Prospective Euromissile Treaty Would Open New Opportunities," *Christian Science Monitor*, September 3, 1987.

Part II

U.S.–Soviet Relations

7

Détente Is Not Enough:
For a New Model of
Soviet–American Relations

Sergei M. Rogov

The state of Soviet–American relations is of paramount importance to international affairs in the contemporary world. If this relationship is stable and constructive, then our interdependent civilization has a good chance of surviving the nuclear age and the challenges of other global problems which have been created by the gap between rapid technological improvements and slower social progress. If the United States and the Soviet Union enter into a nuclear conflict, they not only destroy each other but also society all over the world, at least as we know it in its present form. Even if nuclear suicide is avoided, high levels of tension and the preoccupation with the nuclear arms race still persist, and the resolution of pressing problems of ecology, raw materials, poverty, starvation, and diseases will be tremendously complicated. The resolution of these problems is vital for civilization's survival into the twenty-first century and beyond.

Is a normal, stable relationship between the U.S.S.R. and the U.S. possible? Sometimes it is alleged that they are doomed for confrontation, but this point of view is not supported by history.

There have been several periods in the more than 70 years of U.S.–Soviet relations. In my opinion, four models have been historically defined:

- The Hot War — U.S. participation in the foreign military intervention against Soviet Russia after the October Revolution of 1917.

- The Cold War — "nonrecognition" in the 1920s and early 1930s, the Cold War from 1945 until the early 1970s, and the period of tensions from the late 1970s until the mid-1980s.

- Détente — the establishment of diplomatic relations in the 1930s and the détente period of the 1970s.

- The military and political alliance from 1941–1945.

These four historical examples prove that all types of relationships are possible between the Soviet Union and the United States, ranging from a direct military confrontation to a substantial level of coordination in a joint fight against common enemies. A Soviet–American conflict, therefore, is not predetermined. Everything depends on a specific type of interaction of interests of the two states at a given period of time.

There are many definitions of "interests." In my opinion, interests reflect the objective needs that are vital for the survival of a society. They are formulated in the political process and thus subjectively determine specific goals and reasons for actions. This explains both continuity and change in the notion of interests.

The interests of such different countries as the U.S.S.R. and the U.S. are not a mirror image of each other. Both have different interests, some of which do not coincide. But on the whole, Soviet–American relations develop dialectically and represent unity and confrontation, competition and cooperation on various types of interacting interests:

- Conflicting interests are aimed at diametrically opposed goals, which mutually deny each other and exclude the possibility of compromise.

- Diverging interests compete with each other, pursuing very different goals while not directly clashing, thus making possible some *modus vivendi*.

- Parallel interests reflect very similar purposes, but each side may try to achieve them unilaterally, without much cooperation from the other side.

- Common interests imply the existence of coinciding goals, which cannot be reached unilaterally and which demand a high level of joint action.

All of these types of interacting interests are present simultaneously in Soviet–American relations at any given moment, but receive different priorities. These priorities, in turn, determine which model of relations the two nations have. When conflicting interests played the main role, it led to a military confrontation. The Cold War relationship appeared when diverging, but not mutually exclusive, interests became overwhelming. Détente began when parallel interests were in the forefront. When priority was given to joint, common interests, a military and political coalition was created.

Balance of Interests vs. Balance of Forces

Obsession with military power has dominated the U.S. approach toward the Soviet Union during practically the entire post-World War II period. The conditions of U.S. nuclear monopoly and later, nuclear supremacy, were seen in Washington as a natural situation. The arms race and creation of a favorable balance of power for the U.S. have become not just a continuation of, but a substitute for, American policy. For the White House, force was not only a means but an end in itself.

The U.S. forced "the rules of the game" on the U.S.S.R., which had to react accordingly. This chain of action and reaction (sometimes overreaction) created the conditions for "mutually assured destruction." Although the continuation of the arms race became meaningless for the United States since the Soviet Union created an equal threat of nuclear destruction for American territory, another revival of the balance of power policies in Washington occurred at the end of the 1970s and the first half of the 1980s. This time, the Reagan Administration counted on not only the economic and technological advantages of the U.S. and its NATO allies, but also on the noticeable evidence of stagnation and pre-crisis developments in Soviet society to help achieve superiority.

Yet, these actions have not achieved their intended goal. The U.S. failed to change the military balance in its favor, and *perestroika* in the Soviet Union demonstrated that socialism's potential is far from being exhausted. At the same time, the arms race has turned out to be a burden for the economies of both countries. The United States today is not omnipotent. The national debt has reached enormous proportions — almost three trillion dollars — and has become the number one national problem. Added to this is another problem — the low competitiveness of the U.S. in the world markets. The gap between the aims and the means of power policies has become obvious.

Today, the Soviet Union counterposes the concept of balance of interests and mutual security with the policy of force and a militaristic doctrine. Recent Soviet initiatives opened the way out of the dead-end which had been created by the policy of confrontation. Soviet–American relations have lately passed their lowest point and are acquiring a more dynamic character. The summits in Geneva, Reykjavik, Washington, and Moscow allowed for the first time in many years the beginning of a real dialogue which includes the entire range of interests of the two countries. The first agreement in history on elimination of two classes of nuclear weaponry (intermediate- and shorter-range missiles) is a practical result of these talks. There are now real prospects for an early conclusion of the treaty on the 50 percent reductions of strategic nuclear arms, if both sides agree to follow obligations under the ABM Treaty.

But it is too early to claim that the Soviet–American relationship has fundamentally changed. This can be achieved only within the solid framework of a balance of interests between the two countries. Such balance demands mutual compromises and agreements beneficial to both sides, which means recognition of not only one's own national interests, but also the legitimate interests of the partner. This balance of interests does not imply a complete end to any competition. The differences in the ideological and socioeconomic character of the two states make some competition unavoidable. But first, the competition must be conducted only by peaceful means. Second, one should not overestimate the Soviet–American contradictions and see them only from the point of view as a struggle between the two systems.

Priority should be given to common and parallel interests, while conflicting and diverging interests should be substantially limited.

The Need for a New Political Thinking

This new approach demands an end to ideological myths and propagandistic stereotypes which present the other side as "an absolute evil" and giving up the outdated "enemy image." It is necessary to recognize the right of every people to have an independent choice of their way of life, the right to have a revolution, or to maintain the status quo. It is also necessary to understand unavoidable differences between nations, societies, and states, and to see them as something positive, as a source for development and competition.

As far as ideological interaction is concerned, it should not be seen as "the war of ideas." On the one hand, an ideological debate is impossible to solve by the power of weapons, especially nuclear weapons. Consequently, the competition of ideas presupposes the methods of a civilized dialogue — persuasion and not an appeal to the arguments of "a big stick." Ideological debate should not be transferred into the sphere of interstate relations.

On the other hand, the content of an ideology is not limited to class views only. The hierarchy of ideological values inevitably includes general (common) human aspects. It should be remembered that Marxism–Leninism did not come from nowhere but was a product of the mainstream of the world culture and relied on many historical achievements of the human mind. The ideology of scientific socialism is a continuation of the humanistic tradition and concentrates on the problem of overcoming human alienation.

Thus, the ideological hierarchies of "East" and "West" share many common foundations, among them humanistic values. It is the higher levels of ideological hierarchies which are occupied by conflicting, mutually denying, ideological values that represent views of different social classes. That gives a confrontational character to the entire mechanism of the ideological interaction. From that sphere, the confrontation spreads into other areas, including political, military, and economic spheres.

The problem, then, is how to control ideological competition and how to prevent it from being projected into the whole Soviet–American relationship. In my opinion, that also demands new political thinking, which leads to less confrontational ideological interaction.

Contemporary civilization is one despite the existence of two competing socioeconomic systems. Nowadays, more than ever, people feel their interdependence

— their unity before the unprecedented challenges which threaten the biological existence of the human race. That is why there is a common interest in survival, which is universally getting top priority and is making everything else subordinate to it. The priority of survival over all other ideological values in the nuclear and space age does not demand a complete rejection of class values, but it deprives the ideological interaction of confrontational absolutism. It creates, besides the ideological poles of repulsion, areas of agreement, mutual exchange, and enrichment by common human values.

The top priority of mankind's survival creates a new shared level of values in ideological hierarchies. Thus, the ideological superstructures will have both a common foundation and "a common roof." Conflicting class values will not disappear, they will stay. But they will be controlled both from "the bottom" and from "the top" by common human values. Competition will continue, but it will be accompanied by cooperation. Thus, the ideological sphere will not destabilize the whole range of Soviet–American relations.

The Paradox of Security Interests

Paradoxically, common interests are most obviously displayed in the area of security. This area, which traditionally served as the source of conflict, today is very convincingly discrediting the one-sided approach based on force. Moreover, it is suicidal to try to build one's security at the expense of the other side. The reason is that civilization has, for the first time in its history, created the means for its own self-destruction. As Mikhail Gorbachev and Ronald Reagan concluded in Geneva, a nuclear war between the Soviet Union and the United States cannot be fought and won. There are no rational political goals which can be achieved through the use of nuclear arsenals. The nuclear armageddon would mean mutual annihilation and the end of the civilization on earth as a result of nuclear winter and global radiation contamination.

That is why balance of forces policy based on nuclear weapons has no prospects for success. There is no truth to the notion of some American "civilian strategists" that the relationship between the U.S.S.R. and the U.S. is a "zero-sum game" where gain on one side automatically equals loss for the other side, and vice versa. The continuation of the arms race makes both sides lose, because the more force one accumulates, the greater the threat is to its own security as a result of the other side's reaction. The reality of this vicious chain of action and reaction in the nuclear arms race makes the United States and the Soviet Union less and less secure. The present level of force imbalance is irrationally high, creating an equal insecurity for each side instead of an equal security. The policy "from a position of strength" leads to a situation where parity means growth. The threat is becoming greater for both sides, and strategic stability is turning into the stability of the arms race. Thus, parity is losing its role as a military and political deterrence factor.

Today, the security of the U.S.S.R. and U.S. cannot be achieved by means of military technology. A rational approach to the military area requires reasonable sufficiency, not superiority. The level of sufficiency is determined by the composition and structure of military forces required for defensive, not offensive, operations. The Soviet Union stands for the complete elimination of the most threatening means of mass destruction, nuclear weapons. Although in the West the concept of the nuclear deterrence is still dominant, many recognize the necessity of stopping the arms race and cutting most destabilizing weapons. According to estimates of Soviet scientists, it is possible to reduce the present level of nuclear armaments in several stages by 95 percent without threatening the security of both sides. Conventional forces and armaments should be also cut to the minimal level necessary for purely defensive purposes.

Soviet and American security can only be mutual. Is it not indicative that the two great powers, which have created tremendous potentials for destruction and are capable of annihilating each other many times, are presently discussing quantitative and qualitative characteristics of their armaments for the purpose of their reduction and elimination? They are also exchanging very sensitive information, which is denied to most of their own citizenry.

Already one can envision a future mechanism for the coordination of interests of both sides in the sphere of security. The Standing Consultative Commission, the Geneva framework for negotiations on nuclear and space weapons, the nuclear risk reduction centers in Moscow and Washington, the on-site inspections of the INF Treaty, the regular meetings between the Soviet Foreign Minister and the Secretary of State, and the growing contacts between the military, may all be seen as "the building blocks," the prototypes, for the forthcoming stucture of mutual security. Through this bilateral mechanism, the U.S.S.R. and the U.S. will have come step by step to a joint understanding on the level of armaments sufficient for each side's legitimate defense needs. This process, in my opinion, will strengthen stability, eliminate "the trust gap," and establish reliable conditions for transition to a nuclear-free, nonviolent world.

Economic Interests

Security (war prevention) is not the only area for cooperation. The U.S.S.R. and the U.S. have common and parallel interests toward many contemporary global problems, such as the ecological crisis. Until now, each country has been trying to solve environmental problems in an independent, or rather parallel, way. It is obvious, however, that efforts in this sphere (as in space exploration, fighting against hunger, and epidemics including AIDS) should be coordinated and combined in the interests of American and Soviet people and in the interests of the whole world.

I would like to especially point out the anachronism of economic warfare, conducted for many years by the United States and its allies against the Soviet Union. In an interdependent world, to deliberately undermine other economies adds to your own economic problems. *Perestroika* in the Soviet Union and the faster growth rate of our economy does not threaten anyone; just the opposite is true. If we are more successful in the realization of our economic potential, then we can make a greater contribution to the stability and progress of the global economy and to the expansion of the world market. It is obvious that the Jackson–Vanik and Stevenson Amendments and other discriminatory measures against the Soviet Union are of ideological, not economic, origin.

I would also like to point out the often forgotten fact that the economic interests of the U.S.S.R. and the U.S. do not clash directly. Historically, economic contradictions were the source of political and military conflicts. Soviet–American competition has been unique, because it is not determined by economic interests. Today, it is impossible to deny that the main economic competitor of the United States is not the Soviet Union but Japan. The U.S. also competes with Western European countries. While looking for active participation in global economic relations, the U.S.S.R. is not trying to "push out" the United States, to disrupt American economic interests in different parts of the world, or to deny Americans access to markets and raw materials. The Soviet program of international economic security rejects neocolonialist methods, but recognizes historically developed, legitimate economic links.

Trade and economic relations between the U.S.S.R. and the U.S. until now have been of secondary importance for both countries. Certainly, there are real opportunities to expand the present volume of trade several times. There is great potential for joint enterprises. Objectively, the development of economic ties can eventually be mutually beneficial. There is also a clear and vital common interest in freeing both economies from the burden of the arms race. The real economic priorities in times of *perestroika* in the U.S.S.R. and the budget deficit in the U.S. demand an early elimination of wasted money and intellectual efforts that are annually invested by the United States and the Soviet Union into the development and production of weapons.

Political (Geostrategic) Interests

The United States and the Soviet Union have political interests of a global nature. That does not give them any "special rights," but provides them with great responsibilities, which can be either a source of tensions or a foundation for international peace and stability.

The most important geostrategic interests of the U.S. and the U.S.S.R. are connected with their allies in Europe (NATO and the Warsaw Pact). It would be very dangerous to threaten these vital interests, especially under the conditions of an unprecedented concentration of military forces in Central Europe. The Soviet Union is proposing the concept of the "common European home," which recognizes existing political and economic ties, but provides for deep asymmetrical reductions in Warsaw Pact and NATO forces and dismantling the system of military confrontation. That is not meant to "decouple" Americans and Western Europeans. The United States and Canada are legitimate participants of the so-called European process. The goal of the common European home can be achieved only if both the West and the East possess enough political cohesion to make brave decisions needed to end the Cold War. That should be clearly understood by both sides to avoid any misconceptions.

Regional conflicts have been an important source of instability in Soviet–American relationships. Unfortunately, as post-war history demonstrated, the sharp clashes of political interests of the U.S.S.R. and the U.S. were usually connected with complicated problems of the Third World. Soviet–American differences were often projected into regional situations. Almost all regional conflicts (except the Falklands War and the Iran–Iraq War) have turned into Soviet–American confrontations, bringing the world several times to the brink of a nuclear holocaust. It is well known that regional conflicts had quite a negative impact on détente in the 1970s, serving as a pretext to undermine the arms control process and the development of bilateral relations.

Today, new political thinking demands a new approach to regional hotbeds of tension. It is becoming more and more clear that in the conditions of nuclear stalemate on the strategic level, not only a nuclear war, but any war (of greater scale than the invasion of Grenada) cannot serve as a means for a rational policy. The conflicts in the Middle East and Indo-China, Central America and the Persian Gulf, South Africa and around Afghanistan have proven the futility of military solutions to political problems.

Each people has the right to choose its own way. This choice, however, should not be abused and converted into a fratricidal blood bath that continues for years or even decades. That is why, according to Mikhail Gorbachev, the new political thinking is expressed by the policy of national reconciliation. This trend is evident in the resumption of political dialogue in Afghanistan, Kampuchea, and Nicaragua.

At the same time, there is a need for "rules of behavior" for the superpowers that would prohibit any intervention into the internal affairs of developing nations and preclude a direct Soviet–American confrontation. That demands a refusal to use, or threaten to use, military force and subversive activities against Third World countries, a refusal to export revolution or counterrevolution. At the same time, the U.S.S.R. and the U.S. may play an important positive role, facilitating the process of peaceful settlement. They cannot impose peace, but can serve as guarantor for a political solution, for instance, under the United Nations' auspices. The Afghanistan Agreements, signed April 14, 1988, by the Soviet Union and the United States, may establish a precedent for such constructive cooperation.

Toward a Cooperative Model of the Soviet–American Relationship

Analysis of contemporary interests of the Soviet Union and the United States shows that there are prerequisites for a solid balance of interests between the two states. Such balance demands material, political, legal, moral, and psychological guarantees; the creation of a mechanism for mutual assessment and account of interests; and development of cooperation in different areas. That, of course, is not a euphemism for the bogey of the "condominium of the superpowers" to dictate their will to the whole world. These fears are unfounded, to put it mildly, because the differences between the U.S.S.R. and the U.S. are not going to disappear; the competition between them will stay. But they will make an indispensable contribution to the solution of universal global problems by dealing with the issue of strengthening strategic stability, organizing the process of nuclear disarmament, and cooperating in the settlement of regional conflicts. Without better Soviet–American relationships, it will simply be impossible to normalize international relations as a whole and establish the mechanism for general security.

Although the model of détente is preferable to the Cold War, détente is not enough for this purpose. A cooperative model is needed to make relations between the U.S. and the U.S.S.R. enduring, stable, and positive.

As Mr. Gorbachev said at the meeting with a group of American senators and scholars on March 11, 1988, "The duty of both nations is to find an entrance to the solid road of constructive interaction, the balance of interests, to learn how to live together in the contemporary complicated world and to cooperate for their own benefit and the benefit of others."

Both countries nowadays face difficult problems in their domestic agenda. That requires a sober assessment of one's own possibilities and interests, and a realistic approach to foreign policy. *Perestroika*, a qualitatively new stage in the life of Soviet society, has already begun in our country. There have recently been some shifts towards realism in U.S. political life too. If the phases of internal development in both countries coincide, then the building of the Soviet–American relationship will have a firm foundation. This will finally allow an end to the post-war period of confrontation between the U.S. and the U.S.S.R. and a move toward the model of cooperative interaction of the Soviet Union and the United States and of all mankind on the basis of the balance of interests of all sides. This model provides for cooperation in radical arms reductions, peaceful settlement of regional conflicts, a solution to the ecology problem and other global issues, development of mutually beneficial economic ties, and organizing an open dialogue and broad exchanges in the humanitarian sphere.

8

The Future of Soviet–American Relations: Preserving the Fruits of the Cold War*

Jerry F. Hough

Nothing is more complicated than to predict the Soviet–American relationship during the next U.S. presidential administration. A great many variables are involved: the impact of the last three years on American public opinion about Soviet–American relations, the nature of Soviet policy in the future, and the character of unexpected events in other parts of the world (and there certainly will be such events, and, judging from the experience of the past, they will come in truly unexpected places).

More basically, we are leaving the postwar period and entering a new world whose outlines and rules of the game are quite uncertain. In the past, Soviet–American relations were actually not that different in periods of détente than in periods of tension such as 1980–1981, and the analyst could be fairly confident that future change would take place within relatively narrow parameters. That is no longer true, and, instead, we can be confident that the world will be very different 25 years from now, but with the precise nature of the change and its tempo still undefined. Yet, if the open nature of the future makes prediction much more difficult, it does mean that we are now in a rare period in which the future can be shaped as well as predicted.

Postwar History

Glasnost has opened many historical questions to discussion in the Soviet Union for the first time, and a Sovietologist watches the process with exhilaration, but also occasionally with dismay; exhilaration, because we now may be able to write the history of the postwar world with the help of far better data and analysis from the Soviet side; occasional dismay, because of a fear that one-sided historical writing of one type may be replaced by equally one-sided historical writing of another type.

The history of the postwar period is extraordinarily complex, and the policies pursued by different leaders and different advisers were far more contradictory than appeared on the surface. Andrei Vyshinsky, whose role in Soviet history was often very unattractive, was a man who advocated a Bukharin–like domestic and foreign policy in *Pravda* in April 1945.[1] Georgii Malenkov, who had an unsavory part in the Leningrad affair, seems to have been a very innovative thinker in foreign policy.[2] Nikita Khrushchev, who could have transformed the postwar world with his major troop reductions between 1955 and 1960, destroyed his own policy with his ultimatums over Berlin. And while Henry Kissinger is often described as imaginative and Andrei Gromyko as stodgy, it is at least possible that it was Gromyko rather than Kissinger whose policy had the subtle sophistication of a Metternich.

The complexity of the postwar period goes beyond the policies advocated by individual persons. It extends to the characterization of the entire period. Most Americans see it as characterized by all-out Soviet–American hostility, moderated only by modest arms control agreements, while Mao Tse-tung spoke of a Soviet–American alliance. Both views are extreme, but Mao's was far closer to the truth, at least if we speak about Soviet–American policy in the industrial world.

The problem of characterization also involves the very language that we use. We speak of international relations as the sum of the foreign policies of the world's governments, but the *relations* between two peoples can be very different than the sum of the foreign *policies* of their two governments. When we talk about Soviet foreign policy, for example, we speak about Soviet policy towards arms control, towards China, towards the Middle East, and so forth, but for me a far more important symbol of the Soviet foreign relations is that Moscow (let alone other Soviet cities) does not have a single French or Italian restaurant.

In my opinion, the best way to summarize the postwar period (admittedly in an oversimplified manner) is to say that it featured very abnormal *relations* of the Russian and the American peoples with the outside world (including with each other) combined with a rather cooperative *policy* in the industrial world aimed at preserving the fruits of World War II. The great paradox of the postwar period is that belligerent language and surface hostility served both purposes rather well.

While Americans have often talked about Soviet expansionism in the postwar world, it actually would be more accurate to talk about Soviet isolationism in that period, especially if we exclude the years 1945–1946. Americans have often talked about containing the Soviet Union, but their talk about an Iron Curtain was more accurate in suggesting that the Soviet Union was containing itself. One can argue about the reasons for this relationship, but by the mid-1930s, domestic policy was aimed at isolating the Soviet people from twentith century Western culture and especially from Western market forces. But whatever the cause of policy, one should not minimize the policy's appeal for a broad stratum of people who were insecure newcomers to urban life and whose fears were strengthened by the Great Depression of the West in the 1930s.

The American foreign relations were more complex. On the one hand, American economic institutions had the self-interest to export capital and to sell a broad range of products abroad (including the products of American culture), and this meant that American ideas, culture, and language shaped the world in a way that the Russian could not because of its different economic institutions and isolationist policy. On this level, the United States was far more expansionist than the Soviet Union, but its military-industrial complex played a secondary role in this process.

On the other hand, the American people, like the Russian people, entered the postwar period with great insecurities. George Washington had warned against entangling alliances, in significant part because he knew that involvement in Europe's conflicts could tear apart a nation trying to integrate European immigrants into a single American people.[3] American entry into World War I did, in fact, unleash strong feelings against the German-Americans, and then the interventionist-isolationist debate of the 1930s was essentially an ethnic one (primarily English-Americans and Jewish-Americans against German-Americans and to a lesser extent Irish-Americans). Franklin Roosevelt manifested his genius as a wartime leader when he prevented the domestic tensions of World War I by appointing men with pronounced German-American names to two of the most important military posts: Dwight Eisenhower as commander of Allied Forces in Europe and Chester Nimitz as commander of the fleet in the Pacific.[4] But after the war, Americans still instinctively feared international relations as something that tore at the American social fabric, and the United States became a world power in 1945 with little experience with the normal give-and-take of superpower interaction.

In addition, the fact that America's first major adversary presented an ideological challenge produced a special feeling of insecurity. Americans come from countries with very different languages and political cultures. Immigrants and their children may be unconsciously insecure in their new ideology, while Americans of older stock may feel insecure that new immigration will swamp the existing culture.

Americans even saw the social democracy of Europe as un-American, and they feared the ideological challenge of communism in a way that is difficult to believe in retrospect. We now look back on the early postwar period as one in which the United States towered over a war-devastated world, but many Americans of the time believed that only treason by domestic Communists could explain why the United States was failing so much. (McCarthyism, it should be recalled, did not focus on any supposed Soviet military threat, but on the danger from American Communists.)

Thus, the postwar period began with two regional powers of 1939 suddenly becoming superpowers in an age with totally new nuclear weapons and each with a population afflicted by a high degree of insecurity. The insecurities on each side went far beyond anything created by the military policy of the other side's government, and it is difficult to imagine what plausible Soviet or American policy could have prevented a very tense relationship from developing.

Postwar Geostrategic Stability

As one looks back at the postwar era in retrospect, however, it is striking how stable it was, at least in the industrial world. Compare it with the earlier period. Russia was allied with Germany in the 1880s and then was allied against it from the 1890s until 1917. In 1922, Russia began close cooperation with Germany at Rapallo, but from 1934 to 1939, it was again in an anti-German alliance with France. This policy was reversed from 1939 to 1941 and then reversed once more by Germany in June 1941. Many other changes occurred in the alliance system, and, of course, the boundaries of European countries changed repeatedly in the decades before World War II. In the last 40 years, there have been no changes in boundaries or in alignments at all in the industrial world.

Basically, we take for granted the postwar geostrategic stability of the industrial world, or perhaps attribute it to the central ideological conflict between capitalism and socialism. However, the intense ideological conflict of the 1920s and the 1930s did not have a similar effect, and in my opinion, we should focus more on the geostrategic stability as a defining characteristic of the postwar period.

It seems to me most useful to think of the postwar period through the prism of Ernest May's study of the lessons of history. Generals often plan for the last war, and May insisted that civilians too are also heavily influenced by the decisive events of their youth and that those are the particular "lessons of history" that they learn (and overlearn).

What would one predict would be the major lessons of the past that would affect the leaders of the postwar? In fact, two generations dominated the period. The first was the Stalin–Molotov–Truman–Eisenhower generation that was born around 1890. The second was the generation born 15 to 20 years later — the Brezhnev–Gromyko–Kennedy–Reagan generation.

For the older of these two generations, the chief lesson of history was that conflicts between England, France, and Germany had destroyed their lives in their twenties and then once again in their forties (and, as has been seen, this included political stability in the United States). The younger generation remembered only the origins of World War II, but it remembered it well. Brezhnev was 35 in 1941, Gromyko was 32, Reagan was 30, and Kennedy, 24.

If one looks at the postwar period in retrospect, it is striking how it solved the problems of the youth of the two generations that ruled the postwar world. For 40 years, the Soviet Union had sufficient control of Eastern Europe to prevent the rise of the type of conflict that was the spark of World War I. The United States was accepted as military leader by Western Europe and Japan, and the belief that a common defense was needed against the Soviet Union ended the conflicts within Western Europe that had led to both world wars. Despite Western anxiety that Soviet conventional superiority and the installation of SS–20 missiles might Finlandize Europe, they actually had the effect that the realist theory of international relations predicted: the perception of a threat was met by a counter-alliance that has been the longest-lasting military alliance in history.

The postwar period was successful in other respects as well. The United States has very few Russian-Americans, and the conflict with the Soviet Union created few domestic tensions. Instead, the muting of conflicts within Western Europe welded together a wide range of European-Americans.

One can argue whether the consequences of the postwar period were intended or not. I personally think that many in the older generations were quite conscious of what they were doing, even if they seldom spoke about it in public. NATO began with a French demand that the United States pledge to keep forces in Europe as a condition for French withdrawal from its zone of occupation in Germany. The French made it clear that they were not seeking American protection against the Russians as against the Germans. Konrad Adenauer was a Francophone from the Rhineland who had always been deeply suspicious of Prussia, and he very much wanted German–French reconciliation. He, too, saw American involvement as very useful.

Similarly, I suspect that many of the older Soviet leaders were quite conscious of their priorities. In 1944, Stalin told a visiting Yugoslavian delegation that he anticipated another war with Germany in 20 years.[5] His determination to divide Germany surely was a conscious one, and he may have had an even more Machiavellian policy of deliberately trying to maintain American control of Europe.[6]

Certainly there was an interesting contradiction in Stalin's postwar ideology. He was extremely insistent that capitalist governments were subordinate to the monopolists of their respective countries and that wars among capitalist countries were still inevitable for economic reasons. Yet, he also insisted that the United States was totally dominant in the capitalist world, even over governments such as Nehru's in India and Mossadegh's in Iran that were claiming to be neutral and even anti-Western.[7] But if capitalist governments were subordinated to their respective monopolists and economic factors were strong enough to drive capitalist countries to war with each other, why would they accept American domination? And if the United States, for whatever reason, had achieved dominance in the capitalist world, how could war among capitalist countries be inevitable?

It is possible that Stalin really was saying that the capitalist countries, left alone, would inevitably go to war as they had in 1914 and 1939 and that American dominance was desirable to avoid a repetition. When Stalin rejected all advice to stay in the United Nations to veto United Nations intervention in the Korean War in 1950,[8] it is at least possible that he saw the intervention as useful.

Similarly, the policies followed by Andrei Gromyko had the effect of strengthening the Western alliance. Some consider this a sign of incompetence, but few men serve 28 years as foreign minister if their policies do not achieve the goals that they and their political leaders desire. All of the public talk about "German revanchism" seemed to be a defense of a conscious policy. It suggested that an FRG independent of NATO would destabilize the GDR, that it would ultimately increase the pressure for the FRG to obtain nuclear weapons, and that it would increase the possibility of dangerous intra-European crisis.[9] And once the Soviet Union acquired intercontinental rockets, and Western Europe lost its military significance in a Soviet–American war, then this policy became not only a response to the "lessons of the past," but also the correct policy for the Soviet Union from a geostrategic point of view, for the reasons indicated.

The Future Environment

Any understanding of future American policy towards the Soviet Union must begin, I think, with a realistic understanding of the postwar era. Basically, there has been 40 years of peace in Europe at the cost of a low percentage of American GNP, and, at home, the policy has not only ended conflicts among Americans from Western Europe, but has also pleased the many Americans with roots in Eastern Europe with bad memories of various Russian policies of the past. Anti-communism has proved a useful tool in speeding the assimilation of American political values among the waves of new emigrants, many from outside Europe.

As a consequence, when faced with the thought of a transformation of the postwar world, American policymakers have instinctively — and sometimes consciously — reacted in accordance with the American saying, "If it ain't broke, don't fix it." They want Soviet–American dialogue and a little meaningless nuclear arms control to keep the population reassured. They are happy enough with a good atmosphere in Soviet–American relations, although some leading theorists worry about that. But deep in their heart, they want a continuation of the status quo because it has been quite satisfactory.

American policymakers also have an instinctive — and sometimes conscious — sense that change will complicate their lives. An end to the perception of a Soviet threat will make West–West relations more difficult to manage. If Soviet economic reform and integration into the world economy leads to a greater economic integration of Poland, Latvia, and the Ukraine into Europe, they will change the attitudes of Polish-Americans, Latvian-Americans, and Ukrainian-Americans towards the Soviet Union and may cause a conflict with those Jewish-Americans who place a higher priority on emigration. This will make American national relations more difficult to manage.

Similarly, conventional arms control is deeply troubling to policymakers. Even within the arms control community, policymakers have worked out a politically satisfactory compromise. They have given symbolic (and inexpensive) nuclear arms control to those who want it, and they have built up more expensive conventional weapons to satisfy the manufacturers, trade unions, and other constituencies in the defense industry. Significant reduction of conventional arms would disturb the domestic political equilibrium that has been established.

Bureaucrats, Soviet or American, seldom openly oppose change. Instead, they point out obstacles to change. They present analyses that "prove" that change will not be necessary, and, often, this is not conscious disinformation. People convince themselves that something unpleasant will not occur before they try to convince others. Such bureaucrats and policymakers exist in the foreign policy sphere as well as industrial ministries, and they exist in the United States (and elsewhere) as well as in the Soviet Union. As a consequence, most of those in the American foreign policy community have convinced themselves that Soviet economic reform, especially Soviet foreign economic reform, will be very slow and that the United States will be required to make only minor adjustments. They have convinced themselves that the technical difficulty of conventional arms control, coupled with the resistance of the Soviet military, will make substantial conventional arms control impossible, and that the talks will drag on for years.

This attitude is reinforced among sophisticated American foreign policy analysts by the understanding that the Soviet Union has reason to be satisfied with the basic geostrategic status quo. Japan has a GNP that is roughly equal to that of the Soviet Union, but spends only 1 percent of it on defense. If it became independent of the United States and spent the 6 percent of GNP that the United States spends — still a small figure in comparison to the Soviet Union — the Soviet Union would be horrified. In this view, any fundamental military change in Europe, let alone withdrawal of American troops, might destabilize Eastern Europe. As a consequence, these analysts believe that the Soviet Union will pursue a less active foreign policy, but one that is still American centered as in the past, in order to concentrate on domestic reform.

If, in fact, this American consensus view of Soviet foreign policy is correct, then it is not difficult to predict the outlines of American foreign policy, regardless of who is elected president. The United States is ready to go back to the détente of the late 1970s, but one must always remember how limited that détente was. Unless a framework agreement is signed with President Reagan, a START agreement will be extremely difficult to negotiate in the next administration. Former president Reagan and other conservatives will say that the good atmosphere of 1988 was achieved with few American concessions, and they will make it difficult for a new president to go further.

Nevertheless, modest disarmament steps seem possible, a minor loosening of economic restrictions will occur, and Third World issues seem unlikely to be as serious as they were in the 1960s and 1970s. If the consensus American view of Soviet foreign policy is accurate, then it seems likely that the correct and civilized relations that Mikhail Gorbachev advocated in May 1985 seem possible.

One should never forget, however, what the older generation of leaders understood: the Soviet–American alliance of 1945–1988 utilized a considerable amount of tension in the relationship to preserve the justification for NATO and the

Warsaw Pact, and it will be difficult to avoid that in the future as well. Situations such as Afghanistan and Nicaragua permitted the creation of such tension without the danger of major Soviet–American confrontation, and if we solve such regional conflicts, we need to be careful that the tension in the future is not created in more dangerous places.

Another Perspective

The problem with the analysis of the last section is that I believe that the American consensus view of Soviet foreign policy on which it is based is incorrect. Although the postwar era has been satisfactory for the Soviet Union in traditional geostrategic terms, it has been very unsatisfactory in other respects.

First, if there has been a Soviet–American alliance in the industrial world to preserve the fruits of World War II, then the Soviet Union has borne a disproportionate share of the cost — of the financial burden — of that alliance. Once the dangerous tensions within Europe had essentially been resolved by the late 1960s and once the Vietnam War ended in 1972, the only justification for the abnormally high level of expenditures was a domestic political one for Leonid Brezhnev: so long as the West was speaking of Soviet military superiority, it was difficult for opponents to the General Secretary to argue that Soviet power was declining under him, as it was in the last decade of his life. The Soviet leadership, like the Chinese, seems determined to shift resources from the military to modernization, and only a reduction of expenditures on conventional forces will achieve that. Hence, in my opinion, it is not likely that the Soviet Union will tolerate another decade of fruitless conventional negotiations in Vienna.

Second, the current generation of Soviet leaders, and especially the population under the age of 55, is not satisfied with the foreign *relations* of the Soviet Union of the last 60 years. Nearly 55 percent of Soviet citizens have a complete secondary-education or better, and this figure is much higher for those under the age of 40. Since the 1950s, the Soviet youth have wanted Western clothes, popular music, films, and so forth, and now the 20-year-old student of 1955 is the 54-year-old "bureaucrat" of 1989. While, in my opinion, Evgeny Primakov correctly analyzed the reasons for weak support for political reform that seeks a multiparty system or its equivalent, I think that Soviet intelligentsia are severely exaggerating the bureaucratic and worker resistance to other types of revolutionary political and economic reform. Americans have spoken of a Russian drive to warm-water ports; I think that we should now speak of a drive to warm-water beaches on the Riviera, on Capri, and in the Greek Isles by bureaucrats and workers who are very underprivileged in comparison with their counterparts in Western Europe.

In my opinion, advocacy of a "common European home" is not only a statement of foreign policy. It is, first of all, a promise of a change in domestic policy that is indispensable to maintain political stability within a population that has changed qualitatively (as late as 1959, only 14 percent of the population had a complete secondary education or better) and that wants a very different relationship with the outside world than the older generation.

Ultimately, however, it has been Soviet secrecy and isolationism that has been more important in creating a sense of threat in the West than Soviet weapons. There has been a feeling that a country that is so secretive may be hiding sinister or irrational intentions. But now, as barriers between Russia and Western Europe disappear, as the Soviet Union becomes progressively less secretive, the sense of threat in Western Europe will continue to decline. It is difficult to believe that conventional arms control negotiations can be dragged out forever when the failure of those negotiations have the implication of an extension of the draft for West German youth who will be rapidly declining in number in the 1990s.

Third, and most important, Soviet economic autarchy, even in the period of détente, meant total protectionism for Soviet industry — a level of protectionism that makes Japan look like a free-trade economy in comparison. Even when the Soviet Union imported technology, Soviet manufacturers lost no sales and did not have to raise the quality of their goods to meet foreign competition (and that is the chief benefit of importing foreign goods). Unlike the leading Japanese firms, Soviet industry was not forced to export manufactured goods to competitive markets, and hence, unlike the GDR manufacturers who traded in the FRG market, they did not have to raise the quality of their goods to meet foreign competition. The Soviet Union neither permitted foreign investment inside the Soviet Union nor investment abroad, again with the same consequences. The result of the total protectionism is what a Western economic textbook would predict: poor quality production, insufficient innovation, inefficiency, and lack of responsiveness to consumers.

In my opinion, the Soviet leadership understands this and is determined to reverse policy decisively. As soon as the enterprises do not receive 100 percent *goszakazy*, the Soviet Union will follow the policy of the Pacific countries and force its manufacturers to export their low-quality goods (at low prices as the Pacific countries did at first) as a way of forcing an improvement in quality. It will develop world-wide networks of joint production and foreign investment. It will take years to have multinational corporations on the Japanese scale, but it will move rapidly in that direction. Nearly 40 percent of Western trade can take place between American corporations and their foreign branches, and the Soviet Union will move in this direction.

In my opinion, we should not understand these three changes as Soviet desires, but as Soviet imperatives. They are required for domestic political stability, and they are required for foreign policy reasons if the Soviet Union is to retain — some would say, regain — great power status. If the older leadership was determined not to permit economic reform nor a change in Russia's relations with Europe, then it wisely concentrated almost exclusively on the defense of geostrategic interests, traditionally defined, and it did so with sophistication. But in a world in which a new leadership must give other interests greater priority, the question of how to balance these interests with the traditional geostrategic ones will be a vexing one.

Conclusions

If my view of Soviet foreign policy is correct, what will be the American response? What will be the character of the Soviet–American relationship?

These questions are extraordinarily difficult to answer, for the United States as a whole simply cannot imagine as possible a Soviet policy that I think inevitable. Hence, it has not really begun the process of developing a response.

Objectively the type of Soviet policy that I have hypothesized should mean a real improvement in Soviet–American relations. Both retain similar geostrategic interests in the industrial world, but both have had substandard growth and increases in productivity of labor in the 1970s and the 1980s, and both need to devote more resources to investment. A decrease in military spending is an obvious way for the United States to reduce its deficit without increasing taxes. Moreover, as Americans increasingly see economic power as crucial in international relations and Japan as the superpower that is America's most serious adversary, those Americans with an intensive emotional fear of the Soviet Union and Communist ideology will find it easier to keep the Soviet Union in proportion and have a more normal relationship with it.

Subjectively, however, the situation is much more complex. First, as already noted, the United States is nervous about major changes in the relationship with the Soviet Union that may create problems within its alliance system. This is intensified by the sense that the necessary effort to reduce its huge deficits with its allies is going to create strains in the alliance and that political-military changes may make the process more volatile.

Second, as Americans have come to understand the increasing importance of economic power, they have not yet accepted the Japanese assumption that the maximization of economic growth is the chief goal to pursue and that political considerations should not be permitted to interfere. Instead, Americans have been more inclined to try to apply economic power to achieve political ends. Economic sanctions are increasingly applied not only to socialist countries, but also to countries such as South Africa and Panama.

Consequently, when the United States sees the Soviet Union trying to rebuild its power while pursuing foreign and domestic policies that the United States does not like, it has a tendency to want to use its economic power either to prevent Soviet reform or to obtain a change in Soviet policy as a condition for economic relations. The United States remains extremely sensitive on questions of technological transfer and "technological espionage." Joint ownership and joint production of rapidly evolving products present qualitatively different problems of technological-transfer control than the simple sale of technology. Thousands of salesmen and purchasing agents representing individual Soviet firms or ministries are capable of far more technological espionage than a small number of embassy employees. Of course, the problem is not simply one of direct controls: if corporations suspect that they will be affected by future sanctions at any time that the U.S.

government becomes unhappy at some Soviet foreign or domestic policy action, they will avoid involvement.

In addition, the United States still wants to link credit and most-favored nation status to Soviet performance on certain human rights questions, especially Jewish emigration. These restrictions had relatively little meaning at a time when the Soviet Union would not permit foreign investment inside the Soviet Union and when it had little interest in exporting manufactured goods. However, if the Soviet Union is encouraging investment and is promoting exports at prices commensurate with the relatively low quality of its production, the level of tariffs will be a crucial matter.

If the Soviet Union is determined to pursue a more vigorous foreign economic policy, it surely cannot permit the United States to have a veto power over *perestroika*. If the United States is determined to maintain technological controls, the Soviet Union must take measures to break them (or, rather, to limit them to clearly defined military categories). These measures obviously involve a more forthcoming policy vis-a-vis American allies and industrializing Third World countries.

Obviously, even at best, it will take Soviet foreign economic policy a long time to become as vigorous as Japan's, and, hence, the United States will have time to adjust. Hopefully, it will respond positively to the pressure of its allies to gradually relax COCOM restrictions. Hopefully, increased European and Japanese investment and trade in the Soviet Union will lead American businessmen to press successfully for greater governmental deregulation of foreign economic activity in the United States so that American businessmen will be able to compete in the Soviet market. Hopefully, Soviet officials and specialists have learned from the study of countries such as the United States, the FRG, and New Zealand how foreign politicians make tradeoffs between nuclear arms control and conventional disarmament, between their military policy and foreign economic policy vis-a-vis the Soviet Union, in order to reassure different domestic constituencies, and, hopefully, the Soviet Union will take these complexities into account as it pursues its new priorities.[10]

Yet, we should not assume that any transition from one historical period to another will take place painlessly, and we should not assume that American public opinion has adjusted fully to a new era.

Clearly the relations of the American people to the outside world are different from what they were in 1945. The American people have had 40 years to adjust to superpower status and to the existence of nuclear weapons. They are far more conscious of Japan and Iran as countries with conflicting and competing interests, and this makes it easier for Americans to think of international relations in more subtle terms than a struggle for good and evil against a single "focus of evil." The experience of economic sanctions against such countries as Panama has surely reduced the faith in the effectiveness of this weapon.

Nevertheless, the 1980s have not been a decade in which the American people have wanted to face difficult decisions. Candidates who have even suggested the possibility of tax increases to combat the budgetary deficits have done very poorly in presidential primaries and elections. The population has been very satisfied with the Reagan Administration's policy of "standing tall" by building up military forces, of not using them in difficult situations (even in Nicaragua and Lebanon), but only in extremely low-cost actions such as the invasion of Grenada, the air strike against Libya, and a naval war against Iran.

Finally, of course, the improvement of Soviet–American relations, including the adoption of the INF Treaty, took place with minimal American concessions — certainly with very few concessions or change in policy that were visible to the American people. When President Reagan went to Moscow in June 1988, he did not retreat on SDI, he brushed aside Soviet proposals on conventional disarmament, and he said nothing about a change in economic relations. He spoke about human rights at such length and in such terms that he received very different access to Soviet television than British Prime Minister Margaret Thatcher. But his various speeches in Moscow provided the center of reporting in the United States. The American people inevitably have gained the impression that an improvement in Soviet–American relations will involve few difficult choices.

Arthur Schlesinger, Sr. presented the thesis that the United States has a 30-year cycle in this century. The 1900s, the 1930s, and the 1960s were periods of political transformation, while President Coolidge in 1925, President Eisenhower in 1955, and President Reagan in 1985 all expressed the public moods of self-satisfaction. One would expect the 1990s to be another period of transformation. Each of the past periods has featured different issues, and as the generation that was in its twenties during the Vietnam ear is in its fifties in the 1990s, its "lessons of the past" are likely to focus on the need to readjust America's relation to the outside world and to reduce the military and ideological factor in American foreign policy.

For this reason, the 1990s are likely to be a period of transformation of Soviet–American relations, and in a positive direction. Nevertheless, if the cycle theory is correct, the 1988 election should be more analogous to the 1928 election than the 1932 election. (Indeed, both the Democratic and Republican candidates tried to project a managerial image such as Herbert Hoover's rather than a transforming image.) In addition, of course, the 1930s and the 1960s remind us that periods of transformation need not be quiet political periods without contradictions.

As we move through a difficult period of transition, however, the most important thing to remember is that all "lessons of the past" are partial. Sometimes the application of them has good results, sometimes disastrous. The lessons of the Vietnam period are not the sum of what we need to know for a wise foreign policy. As we leave the postwar period, we should not forget the problems — extremely serious problems — that it solved. Soviet analysts in particular, who were unable to read sophisticated Soviet descriptions of the rationale of Soviet policy in the past, should not assume that Soviet policymakers of the past were always following a dogmatic policy just because their language seemed dogmatic at first glance or because their memoirs do not explore motivation in a sophisticated manner.[11]

As we enter a new era, we need to retain the useful memories of the older generation as well as new insights. We need the benefits of a more normal relation of Americans and Russians to the outside world and of a greater Soviet integration into the world economy, but we also need to retain the positive aspects of the postwar period. No one needs a return to the multipolar politics of the type that existed between 1900 and 1940. No one — least of all a Socialist world becoming part of the world economic system — needs an economic depression produced by economic war between the capitalist industrial giants. A successful transition from the postwar world will require the highest degree of sophistication by both Soviet and American policymakers.

Notes

1. See his two-part article in *Pravda,* April 21–22, 1945. The hard-line answer to Vyshinsky was provided by Peter Pospelov, the chief editor of *Pravda*, in a two-part article published in *Pravda* April 22–23.

2. See, for example, the remarkable article by E. Burdzhalov in the issue of *Kommunist* that carried the obituary of Stalin (no. 4, 1953). Burdzhalov, then head of the party history department of the Higher Party School and working closely with Malenkov, wrote the first article in this issue ostensibly praising Stalin, but essentially reversing his foreign and (implicitly) domestic policy and even suggesting that Stalin's last ideological work was incorrect Menshevism. The table of contents of this issue of *Kommunist,* which was signed to press while Malenkov was both Central Committee secretary and chairman of the Council of Ministers, was basically a list of the reform areas of the next decade.

3. See Seymour Martin Lipset, *The First New Nation: The United States in Historical and Comparative Perspective* (New York: Basic Books, Inc., 1963), pp. 65–66.

4. The United States and the Soviet Union face enormous difficulties in understanding each other's politics because of the fundamental difference in their nationality problem. For example, Armenians who live in Nagarno–Karabakh insist on retaining their Armenian language and culture, while those who have gone to the United States fully accept that their children and grandchildren will assimilate into the English–language culture. Each of these different patterns create powerful political imperatives, but they are so different that each country can hardly begin to comprehend the other.

5. Milovan Djilas, *Conversations with Stalin.*

6. "Lessons of the past" are likely important in understanding an earlier Stalin decision. Stalin surely remembered that mobilization in 1914 had inexorably led to war, and he learned the lesson so well that he refused to mobilize in June 1941 when he should have.

7. For a discussion of the published Soviet discussions that accompanied the formation of the Stalin orthodoxy on the nature of the bourgeois state and of international relations — and the post-1953 debates through which this orthodoxy was attacked and a new way of thinking was developed, — see Jerry F. Hough, *The Struggle for the Third*

World: Soviet Debates and American Options (Washington, DC: The Brookings Institution, 1986).

8. A.A. Gromyko.

9. For a discussion of the theme of German revanchism in the Soviet press in 1984, see Jerry F. Hough.

10. The Soviet 70-year desire to court and use extreme left-wing forces in Europe — Communist parties until the 1960s and then increasingly anti-nuclear forces — has often complicated its conduct of foreign policy, and these complications will become greater in the new age.

11 Since Andrei Gromyko is chairman of the Presidium of the Supreme Soviet, he must write about foreign relations with care. However, the first sections of his memoirs show that he is capable of memoirs on the level of a Churchill or a Kissinger. He owes one last service to his country and the world to write a much more informal, personal memoir of the foreign policy process and the motivations of the individuals making policy, even if such a memoir cannot be published immediately.

9

Soviet Force Posture:
Dilemmas and Directions*

William E. Odom

T he Soviet General Staff has embarked upon the third revolution in military affairs in its history. The first revolution occurred in the 1920s, the second in the 1950s, and the third dates from the late 1970s and continues to this day. Each has had a major impact on Soviet economic, social, and scientific policies. Each has to date received priority over virtually all other aspects of public policy in the Soviet Union. This primacy of military affairs has become a powerful — perhaps the most powerful — constraint and determinant in Soviet political development.

Each of these revolutions has involved a major change in force development policy. Most states do not build military forces randomly, or just to be in fashion, or purely because of bureaucratic momentum. They build toward some mission, to meet some threat, in accordance with some doctrinal rationale — that is, with purpose. To speak of force development policy, then, is to speak of the rationale for developing specific kinds and sizes of forces. Why have tank divisions instead of infantry divisions? Why ICBMs, IRBMs, and ABMs? Why chemical weapons? And why a particular number of each? Why not more? Why not fewer?

The Soviet Union has generally been very advanced in working out new rationales for force development. Its unclassified military literature is among the richest in the world, which is indicative of the existence of an even more extensive classified analysis. Even a casual familiarity with Soviet military force structure would convince a reader of Soviet military literature that there is a strong causal

*This article originally appeared in Problems of Communism, July–August 1985, Vol. XXXIV, pp. 1–14.

relationship between Soviet force development policy and actual Soviet force building. The kinds and numbers of forces cannot be wholly explained by theories positing "bureaucratic momentum" (according to which no one at the top appears able to control the military-industrial complex) or "action-reaction" (according to which the United States "acts" and the Soviet Union merely "reacts"). Someone with a plan does seem to be in charge, even if that someone does not always succeed in realizing the plan's goals. Traditionally, the Soviet General Staff has played the key role in planning Soviet force building , a role that has grown stronger with time.

Yet, despite all its previous force-building successes, the General Staff today faces several new and troublesome challenges with consequences for nonmilitary areas of policy that are likely to be profound. It has succeeded in designing a doctrine and force structure for nuclear war at all levels — strategic, operational, and tactical — that the United States and its allies would find difficult to defeat. The whole Soviet economy has been giving priority to this task for well over two decades. Although the costs have been high, the yields probably appear worth the price as the Politburo assesses the change in the international correlation of forces over these decades. Nonetheless, like Sisyphus who was condemned in Hades to push a rock up a hill only to see it constantly roll back down, the Soviet General Staff appears to see the rock of its military labor rolling back to the bottom of the hill, presenting the Soviet military with a repetition of the same task: another long-term force building. Not only will this task be costly, but it will probably place nuclear weapons modernization into a secondary role as the competition with the West shifts into areas where the Soviet Union is less well prepared to compete. Moreover, the new force building may deprive Moscow of continued success in the political and propaganda aspects of arms control negotiations by raising questions about the ostensible Soviet rationale for such negotiations.

The fundamental cause of the new task is new technology, namely, microcircuitry, directed energy systems, and genetic engineering. But Soviet success in dealing with these new challenges can be significantly affected by Western policies. They can make it easier or harder for the Soviet General Staff to roll the rock up the hill again. However, as I shall argue, the causal relations between some Western policies and Soviet force development are not what conventional wisdom has suggested; in fact, they are the reverse in some instances. For example, contrary to the view popular in the 1970s, expanded East–West trade is likely to hurt, not help, arms control. Certain kinds of arms control agreements could actually contribute to a more rapid build-up of weapons by helping Moscow out of its force-building dilemmas; some aspects of military competition, particularly in the quality of weapons, can improve the prospects for arms control agreements, as well as increase Western security.

To make the case for the centrality of new technologies in Soviet force development, as well as to gain some historical perspective on the current dilemmas facing the General Staff, it is necessary to review how the Soviets have handled major changes in force development policy on two previous occasions.

The First Military Revolution

To understand Soviet military problems, one must understand Soviet military purposes. At the highest level of generality, there is little reason not to take key aspects of open Soviet military policy at face value. How do the Soviets explain publicly the need for military forces? After all, in 1917, it had been Bolshevik policy to abolish all regular forces and to replace them with a workers' militia. In the Bolsheviks' view, if the army were distinct from the working class, it could be used as an instrument of suppression. However, if worker and soldier became synonymous, the army could hardly be used against the working class.

This policy, to be sure, was quickly reversed in 1918 when civil war broke out in Russia. But when revolution did not spread to the rest of Europe following the Bolshevik victory in the civil war, the Bolsheviks quickly found a new rationale for their regular army or an "army of a new type," as they called it. The Red Army was necessary to defend the fledgling socialist republic against international imperialism. Thus, "socialism" would have to be preserved behind the shield of the Red Army until the international correlation of forces favored the socialist camp. And this shield would be needed until true peace came with the final and decisive victory of socialism over capitalism.[1]

The Soviet definition of peace is unique and incompatible with Western definitions.[2] Defense, in this peculiar Soviet sense, means offense. Peace means the destruction of all nonsocialist states. If that can be accomplished without interstate wars, that is, through internal revolutions, so much the better. The term "coexistence" also does not mean what most people in the West would understand by it. In fact, peaceful coexistence in the Soviet definition is a continuation of the international class struggle by other than direct military means whenever possible. This policy was conceived in the early 1920s as a strategy for avoiding war with the West, which Lenin believed the young regime would lose. It meant building domestic industrial power to support a military establishment that could prevail in a showdown. It meant maintaining "correct" relations with the advanced capitalist states in order to derive the advantages of economic interaction with them. And it meant supporting revolutions and wars of national liberation in what today is called the Third World. In other words, peaceful coexistence was a strategy for irreconcilable struggle, political and military, with capitalism. Peaceful coexistence remains Soviet policy today.[3]

If one takes seriously this political rationale as providing the basis for Soviet military policy, then the force development of the Red Army in the 1920s and 1930s was remarkably logical and even predictable. Its development was guided by an extensive doctrinal review concerned primarily with the implications of new technologies for future war. Aviation, motorization, and chemical weapons had appeared in World War I. They portended, as Red Army theorists pointed out, a less clear distinction between the "front" and the "rear" in war. Bombing of cities, industrial plants, and military forces deep in the rear areas could be expected. Motorized forces could conduct much deeper operations. The new weapons would also require a well-trained officer corps and a literate manpower pool for military recruitment. Moreover, an adequate research and development (R&D) base and an advanced industrial capacity were imperative for the underdeveloped Soviet Union.

The doctrinal review did not go smoothly in the early 1920s. Debates on almost all aspects of military policy were acrimonious and conducted openly. Struggles within the leadership clique affected the review, particularly Leon Trotsky's removal in 1925 as Commissar of War. Mikhail Frunze, his successor, synthesized all views, capped the debate, and established the general directions of Soviet military doctrine in several short pamphlets, including "A Unified Military Doctrine and the Red Army" (1921) and "Front and Rear in Future War" (1925).[4] The goals were not modest. The Soviet Union was committed to build military power that would ensure the eventual success of the Bolshevik revolution throughout the advanced industrialized world.

Soviet actions followed doctrine. The standing Red Army was reduced to about half a million soldiers in active units backed by a large militia force – a policy designed to save manpower in peacetime. The Red Army became a school for literacy. Officer education became a top priority, prompting the establishment of a general staff academy and a host of other measures.[5] In active combat power, the Red Army was allowed to become quite weak. A foreign observer might have concluded that the Soviet regime was quietly disarming itself, a view that would have been as misleading about Soviet military policy as was the contemporary view that the New Economic Policy meant that the Bolsheviks were reintroducing a fullscale market economy. In reality, the Bolshevik leaders were taking a short-term risk in order to have a large, modern military force in the future.

By the mid-1930s, the regular forces were being expanded. New equipment was being produced by Soviet industry and operational doctrine for deep operations had inspired not only the development of new tanks and airborne forces, but also a massive effort to build a modern aviation fleet. Before Stalin's purge of the Red Army's general officer corps in the late 1930s, there had accumulated a fairly large number of trained officers as a result of the military's education policy. But war came faster than Stalin had expected, and many aspects of the long-range military force development program were still incomplete. This should not, however, obscure the essential rationale that guided force development and the cluster of coherent policies that it produced. The policies were impressively perspicacious in retrospect.

Soviet military historians ascribe an orderly and "scientific" character to the process of analytical review in light of new technological, political, and economic realities, followed by doctrinal proclamations, a programmatic building of weapons and forces, and a working out of what they refer to as operational and tactical art. While they impute to this process a tidy causal chain of events that abuses the historical record in a number of ways, their writings do show both a series of actions and a record of policy intention that appear linked by more than chance correlation. In any event, this is the approach that Soviet military leaders have come to believe they should take with regard to the overall task of force development. In fact, they virtually repeated the process after World War II.

Second Military Revolution

In the late 1940s, the Soviet military found itself in a position analogous to what it had faced in the 1920s. The economy was largely destroyed by war, and the number of soldiers under arms was far too large to maintain. The educational level of all ranks was too low for modern technology. And three new technologies — nuclear weapons, rocketry, and cybernetics — appeared to affect fundamentally the nature of future war.

The Soviet response, perhaps not surprisingly, was also analogous to that of the 1920s. In the process of rebuilding the economy, the military sector was given the highest priority. Most of the active duty manpower was demobilized, bringing the force levels down to relatively small numbers. The system of military education was revamped, in order to bring about a long-term upgrading of commissioning schools and military academies. The nature of future war was redefined in light of the new technologies, and the General Staff set to work on changes in operations and tactics necessary to take these technologies into account. The flurry of activity in the late 1940s and throughout the 1950s gave birth to much of the present Soviet doctrine and force structure.

Assuming that the three technologies would change the nature of future war, Soviet theorists considered it essential that military doctrine come to grips with two central effects arising from them: the large firepower that nuclear weapons bring to the battlefield and the great range and accuracy in the delivery of the firepower made possible by rocketry and cybernetics.

Although a great deal has been written about Soviet doctrinal development in the 1950s and 1960s, most Western analysts tended to see Soviet doctrine as developing along lines similar to those evolving in the West in response to the same problems. However, as the Soviets themselves became more explicit and as the nature of Soviet weapons development became clearer, Western analysts began to uncover a quite different picture of that doctrinal evolution.[6] The overall impression one gains from these later interpretations is that of a pragmatic Soviet effort to master the new weapons, to make them usable for strategic, operational, and in some instances even tactical objectives. This is not to say that there was a Soviet desire to use nuclear weapons. Rather, it is to say that the Soviets assumed that these weapons *might* be used. In that event, they wanted to know how to use them purposefully in support of war aims, rather than viscerally in a sort of hopeless retaliation or bluffing deterrence.

The post–World War II phase of Soviet force development policy was based on three key points. First, weapons of mass destruction require that one's own forces be dispersed in order to present few targets worthy of a nuclear strike. The Soviets solved this problem by echeloning their forces, that is, by spreading them out evenly to the rear so that no really large concentrations could be targeted.

Second, breaking through an enemy's defense requires a massing of maneuver forces. The solution to this problem was found in a high speed of attack, requiring the echeloned forces to move forward at 60 to 100 kilometers per day. This causes rapid accumulation of forces at the front, in close contact with the enemy's defense, and thereby permits breakthroughs and allows redispersion by deep operations into the enemy's rear.

Both of these techniques required the abandonment of a number of traditional principles of military art. In particular, the dispersion of forces made it impossible to mass them for a "main effort" in one sector. The high-speed offensive meant either breaking through where opportunities occurred or along an entire front. Placing a concentration of forces forward for a main effort would put them at risk under the enemy's tactical nuclear fire.

Third, this doctrine would remain empty theory until equipment and weapons systems were produced in sufficiently large numbers to make its implementation feasible. Thus, Marshal Vasiliy Sokolovskiy's 1962 volume, *Military Strategy*,[7] was not a statement proclaiming a Soviet capability to implement this kind of combined arms offensive with nuclear weapons and rocketry support. Rather, it was a statement about the technical capabilities of the new weapons, their implications for future war, and some general ideas about how the new problems that they presented could be resolved. Sokolovskiy and his collective of contributors were actually doing what Frunze had done in the mid-1920s. They were providing a synthesis of a plethora of classified discussion, debate, and analysis that had been conducted in military circles in the 1950s. They were distilling it into a text for officer general education about the major directions and problems for Soviet force development in the future.

If one looked at the Soviet force structure as it emerged in the 1970s, one saw a growing inventory of capabilities to implement the doctrine. Therefore, it is not surprising that more and more references appeared in open Soviet military literature to techniques for conducting war, nonnuclear and nuclear, at the tactical, operational, and even strategic levels.[8]

Again, these references do not mean that the Soviets necessarily want to conduct war at the nuclear level. Rather, they indicate that the Soviets realized that such a war could actually be conducted, and that they were gaining the means to do it. Those means encompassed more than just large numbers of small-yield nuclear weapons. They include armor-protected infantry vehicles, artillery carriers, air defense carriers, engineering equipment, tactical rockets, and frontal aviation.

It is also important to note that strategic defense of Soviet territory, very much a part of the Soviet doctrine in Frunze's day, was not abandoned during the post–World War II period. It has remained a strong element in Soviet doctrine and practice. Civil defense, hardening of command and control, air defense, and ABM development have been generously supported. Whether or not these programs would be effective in an all-out war may be a matter of dispute in the West, but there is no doubt that the Soviet leadership has remained willing to commit significant resources to the pursuit of survivability of sufficient command and control, military forces, and industrial capacity and manpower to sustain a long and drawn-out campaign even if nuclear weapons were to be used massively. In other words, Soviet actions strongly suggest that the Soviets believe that they can evade the "assured destruction" imputed in the West to nuclear retaliatory forces.

Thus, in order to understand why we have witnessed in the last two or three decades the largest military build-up in history, we must grasp the doctrinal rationale behind it. The action-reaction theory, the bureaucratic momentum thesis, and other such explanations miss the critical rationale for the build-up. New technologies, military experience, and fundamental policy aims originating in the early years of the Bolshevik regime were its causes. Actions by the West represented constraints, not causes.

Direction of Third Military Revolution

If this is the historical record to date, what about the present dilemmas and future directions for Soviet force development? To answer this question, we must begin by looking for three kinds of evidence. First, are there any new technologies that promise to have major implications for the nature of future war? Second, is there any Soviet doctrinal writing on those implications? And, third, of course, are there weapons developments and organizational changes that have followed from the doctrinal changes?

There is now abundant evidence in all three categories. The new technologies are microcircuitry, directed energy systems (DES), and genetic engineering. Microcircuitry helps make possible what are called "smart weapons," that is, warheads with a variety of conventional energy munitions that are guided to targets with virtually no error, warheads that can seek a target without external assistance and can discriminate between tanks and trucks, and so on. Directed energy systems are also a part of the set of technologies required for these new families of weapons. They make ranging and guidance possible to a degree inconceivable in previous decades. Genetic engineering is less developed for weapons application, and precisely what it may yield in this area is far from clear. What is clear is that drugs for medicinal purposes could also be used for destructive purposes. New genes could be cloned for a variety of effects on the human body, from debilitating to fatal. The Soviets have made a large commitment of resources to genetic engineering, suggesting that the requirements of the Soviet General Staff affect policy in this area as well.

Interest by the Soviet military in the doctrinal implications of all three technologies dates back to the early 1970s, and possibly even earlier. When Marshal Nikolay Ogarkov was promoted to the position of chief of the General Staff in 1977, a number of other senior officers also moved into key positions, officers who were already noted for their writings about changing technologies and warfare. During the past five or six years, Soviet doctrinal writings have shown a concern for exploiting new weapons and technologies. Ogarkov himself published a notable booklet in 1982 that signaled a major shift in direction, although it did not represent a watershed of the kind seen in the 1950s.[9] In this work, Ogarkov spelled out clearly the tasks for future force and doctrinal development, and he chided his fellow officers for being slow to exploit new technologies.

The focal point of this revision of Soviet doctrine is the "theater strategic operation." Ogarkov argues that wars before 1600 tended to be a series of regimental-sized engagements. In the seventeenth century, they became a series of "brigade" operations, in the eighteenth century, "division" operations, and by the close of that century, "army" operations, that is, battles involving simultaneous operations by two or more divisions, thus requiring an army headquarters and staff to coordinate and control the battle. That form of operation continued, in Ogarkov's judgment, through the nineteenth century until the Russo–Japanese war of 1904, when the Russian command controlled two armies simultaneously. This level of command in the Russian and Soviet lexicon is a "front," that is, two or more armies under one commander. By the end of the World War II, the Soviet command structure was managing two fronts simultaneously as a single operation.

The task today and in the foreseeable future, Ogarkov asserts, is to master this multifront operation or theater strategic operation, to use the Soviet terminology.[10] Geographically, it would involve an attack on a front 500 to 750 kilometers wide and proceed about 1,200 kilometers in offensive depth. The current standard Soviet doctrine for a "frontal operation" is 150 to 300 kilometers of frontage carried to a depth of 300 to 600 kilometers into the adversary's territory. Thus, two or three fronts conducting side-by-side offensive actions to at least twice the depth of a front would cover an area roughly 500 to 750 kilometers wide and reach an offensive depth of 1,200 kilometers.

Needless to say, the scale and speed of offensive operations envisioned by Ogarkov are unprecedented, and it is highly doubtful that the Soviet Armed Forces could execute a theater strategic operation of that magnitude today. Thus, Ogarkov's articulation of the direction of force development and doctrine should be seen as an aspiration akin to Frunze's "Front and Rear in Future War" and Marshal Mikhail Tukhachevskiy's "1936 Field Regulations," which set the depth of the offensive operation at 150 kilometers.[11] Both were stating goals in light of technological capabilities. The same was true of Sokolovskiy's 1962 volume. Although somewhat less radical, Ogarkov's little volume nonetheless sets forth incredibly demanding new offensive depths and frontages. Perhaps the geographical scale is less astonishing than the speed. There will be no pause between frontal operations within a theater strategic operation, thus shortening by weeks the time estimated for reaching the 1,200-kilometer depth.[12]

The stimuli for this change in doctrine, of course, are the new technologies that permit, in theory, the communications, control, and accuracy of fire support for operations conducted on such a scale and at such a tempo. Changes in Western military doctrine, as we shall see, are also a factor. Even without these stimuli, however, it is easy to see how a Soviet strategist would seek this speed and scale. For some time, he has had the theater fire support means to strike the "full depth" to the theater — to use Soviet terms — with nuclear means. Soviet short-range ballistic missiles and frontal aviation are sufficient to execute such a fire plan.[13] The longer Soviet ground forces require to penetrate to the full depth of the theater, the less effective those deep-theater missile and air attacks will be because the opposing NATO forces will have time to recover, reorganize, and meet the attack. It is a matter of "leaning into the artillery," which military commanders of maneuver forces have always known is imperative for a successful attack. That is, attacking infantry and tanks try to get as close as possible to friendly artillery falling on enemy positions so as to cover their assault. Heretofore, Soviet ground maneuver forces have not been able to lean into theater fire support to the full depth of the theater. In a sense, Ogarkov's theater strategic operation represents an attempt to achieve precisely this exploitation of deep targeting within the theater.

Although Ogarkov has not spoken as explicitly of strategic defense as he has of combined arms offensive operations, this should not be taken to indicated a lack of Soviet interest in the matter in the current revolution in military doctrine. The new technologies also create new possibilities for active defenses. The assumption by the General Staff that a theater strategic operation can be executed implies a belief in the possibility of riding out and significantly degrading Western nuclear retaliatory attacks. Moreover, this belief has inspired efforts in comprehensive strategic defense, a Soviet strategic defense initiative that appears to exceed the ABM limitations now in effect.[14]

A great deal of organizational change took place while Ogarkov was chief of the Soviet General Staff. Much of it appears to have been directed toward anticipating not just the so-called revolution in military affairs created by nuclear weapons but also the lesser revolution prompted by the latest technologies. To get some idea of the significance of this additional shift in military affairs, we only need to read the 1984 interview with Marshal Ogarkov published in *Krasnaya Zvezda*. Conventional weapons are becoming so efficient and destructive, he said, that a global war in which nuclear weapons would not be used is a possibility.[15]

The trend in the West toward new, nonnuclear weapons has been under way for more than a decade. That trend, combined with shifts in U.S. military doctrine, has clearly had a role in bringing about changes in Soviet operational doctrine. In the mid- and late 1970s, the U.S. defense establishment began to grasp what the Sokolovskiy directions had begot in Soviet doctrine and force structure. NATO forces would be facing deep-echeloned Warsaw Pact forces that would rapidly grind down NATO defenses through successive attacks at the forward edge of the battlefield. The first echelon of Warsaw Pact divisions, and even the second, might be defeated, but as succeeding echelons — in effect, the next army — arrived, the NATO defense would be facing fresh forces while its own were diminished through attrition. The issue for American planners was whether U.S. forces could afford to let the deeply echeloned forces arrive at the front in such a phased fashion. U.S. Army analyses showed that if the fight were carried through long-range targeting against the follow-on echelons as a "deep battle," while another battle occurred at the front, the follow-on forces would not arrive in good shape and the estimated battle outcomes would begin to shift against them. That shift is a function of the effectiveness — accuracy and warhead power — of deep targeting. Tactical nuclear weapons could be used for these purposes, but the concept of echelonment was intended to make forces less vulnerable to nuclear attacks. In contrast, "smart" conventional warheads and bombs promise to be more efficient for such "deep attacks." Moreover, as new NATO ground force weapons — higher-speed tanks, infantry fighting vehicles, and attack helicopters with precision guided weapons — are fielded in sufficient numbers, this change in tactics and operations promises to create opportunities for ground force counterattacks to tactical, and possibly even operational, depths.

As this new doctrine was taking inchoate shape for the U.S. Army and Air Force under the name "AirLand Battle," it galvanized Soviet attention. The Soviet military had to consider whether this doctrine could destroy the synchronization of deep echelonment moving into battle that is necessary to keep the tempo and to exploit breakthroughs. If weapons and forces for AirLand Battle are fielded, current Soviet doctrine is unlikely to cope adequately without changes. This has provided the second set of stimuli for major revisions of Soviet doctrine, and it is thus not surprising that Soviet authors are openly pointing out flaws in Sokolovskiy's concepts of 1962.[16]

One such change had received wide attention in defense circles in Europe and the United States: the so-called Operational Maneuver Group.[17] It is a concept for committing the second-echelon forces across the front much earlier and much deeper. In other words, the scheme seems to be to mass forward earlier to avoid a U.S. deep attack and to carry a Soviet deep attack into NATO's rear. While attractive in theory, this concept puts even greater stress on command and control, synchronization of movements, fire support, air support, and logistics. At the same time, it may offer earlier and more vulnerable targets for the opponent. The forward massing of forces on an axis of the main effort, implicit in the operational maneuver group concept, is precisely what the Sokolovskiy doctrine of echelonment was designed to prevent.[18] The Operational Maneuver Group seems to be a return to traditional offensive doctrine, but on a larger scale and at a faster pace. It is, of course, a part of Ogarkov's larger concept, the theater strategic operation. It should be understood as one technique in such an operation for getting maneuver forces deeper and sooner to lean into the long-range supporting fire.

Although the Soviet General Staff has defined fairly well the directions for future force development, taking into account both technological change and Western doctrine and forces, it still faces a number of dilemmas in attempting to realize these goals. First, it is already evident that the Soviets recognize the shortcomings in the doctrine that had guided 20 years of force development and officer training. Now, the theater strategic operation concept places even greater demands on the Soviet officer corps, demands that probably exceed its already impressive education and training achievements in the postwar decades. This bold and unparalleled concept of operations will undoubtedly require considerable improvements in command and staff training. Is the officer education system up to another dramatic qualitative upgrading after having just gone through perhaps the greatest one in both Russian and Soviet history?

A second dilemma is whether the Soviet scientific and technological base can support the exploitation of the new triad of technologies for military applications. Or will it simply bog down under the demands placed on it by the military? The answer to this question is not yet clear, perhaps not even to the Soviet leadership. The 1960s and 1970s placed heavy demands on Soviet research and development (R&D) capabilities, which were met in no small part by exploiting East–West economic interaction. What Soviet scientists could not develop, they usually could count on the KGB to buy or steal from the West. Today the acquisition of Western technologies is both more difficult and more necessary because of the increasing complexity of new weapons. A Soviet T–62 tank is remarkably simple compared to a T–72 or a T–80 tank to which computers, laser equipment, and more advanced metallurgical construction have been added. In rocketry, aviation, and command and control, the applied technologies are much more costly and complex to develop and involve many more ancillary technologies and products that must be developed, borrowed, or stolen from abroad.

The weapons for the future, that is, those that are highly dependent on the new technologies — microcircuitry and directed energy — can in all probability not be developed in the Soviet Union without extensive access to Western economies and R&D communities. The "smart weapons," or precision guided munitions, which Soviet military analysts see as changing the nature of war, rely on a variety of innovations in the use of these new technologies. Most of the innovations are being made in the West, and moving them into serial production is not easy, even for Western firms. Although espionage may give the Soviet Union access to the new technologies, only extensive cooperative relations with Western firms that have applied them in mass production will allow the Soviet Union to achieve a respectable indigenous production capacity. Moreover, dependence on Western sources of supply, for example on high-grade silicon, is likely to grow as the Soviet Union develops its own production programs.

Third, can the Soviet economy handle the new production demands? This is really a two-fold question. Can the economy meet both qualitative and quantitative requirements for the anticipated force development and modernization? Again, as in the case of the S&T base, the requirements of the 1970s were easier to meet. Future requirements will place higher per-unit costs on industry, and the quality will have to be much higher for many items. According to the dictates of Soviet military science, new technologies cannot have a significant impact on doctrine until sufficient quantities of new weapons are available.

All three of these dilemmas must seem cruel to the Soviet leaders. After a 20-year struggle to get ahead with forces and a doctrine for nuclear weapons and rocketry, they find themselves confronted with a new and analogous struggle to stay ahead. In many categories of forces they have achieved a clear edge. Yet, if NATO now proceeds with modernization programs that results in fielding many systems with the new technologies, those leads may well vanish.

In Leonid Brezhnev's last years and under Yuri Andropov's general secretaryship, it seemed that the Soviet leadership had committed itself fully to undertake yet another major modernization effort, yet another 20-year program.[19] The doctrinal modifications were set forth, and there is as yet no sign that they have been discarded. The rate at which modernization will go forward, however, may well be in question. We will not know this for some time, and the answer will depend to a significant degree on Western policies. Trade policy, arms control policy, and force development policy in the United States and NATO will either complicate these dilemmas for the Soviet General Staff or ease them somewhat. Since conventional assumptions about the causal nexus in each of these policy areas are open to question, some elaboration of this point is essential.

Role of Western Policies

Since 1980, many analysts in the West have been suggesting that U.S. policies were forcing Moscow to review its basic foreign policy premises and to revise significantly its commitment to detente. However, an assessment of Soviet gains from the détente period and the lack of attractive policy alternatives led me to conclude that while détente might offer less today to the Soviets than in the 1970s, it would still be advantageous for the Soviet Union.[20] For one thing, the confluence of Soviet economic needs and a changing political climate in Europe has made it unprofitable for the Soviet Union to continue to outdo the hard-line U.S. policy. For another, the dilemmas in force development policy have made it necessary for Soviet leadership to revive as much East–West trade as possible. Without this trade, the Soviet Union will have neither the S&T base nor the industrial capacity for its preferred force development path.

Détente in the 1970s facilitated Soviet force development through arms control agreements and the arms control process. SALT I did two important things for the Soviet Union. First, it ratified a large Soviet advantage in a number of strategic systems. Second, it closed off U.S. strategic defense programs, giving the Soviets time to catch up in ABM technology. It thus permitted the General Staff not to have to choose between a mix of ICBM and ABM programs and allowed for the accelerated build-up of the ICBM force without having to fear that the United States would deploy ABMs. The Soviets took advantage of the situation to stay well ahead in ICBMs and to catch up and actually deploy the one ABM site around Moscow permitted by the 1972 ABM treaty.

Two American programs, the M–X missile and the B–1 bomber, have both run afoul not because of technical limits in the SALT treaty but because of congressional opposition to the programs, generated in part by the political disputes between proponents and opponents of the SALT process and in part by debates over practicability. In the case of ABM, the United States has not deployed even the one site permitted by the treaty. Yet now, as the Soviet General Staff faces a second postwar modernization program of enormous dimensions, it sees a reviving U.S. interest in strategic missile defense. The Soviet Union's decision in late 1984 to return to the arms negotiations was to be expected, and the primary Soviet goal to defeat the US interest in ballistic missile and space-based defenses as effectively as it had defeated ABM in 1972.[21]

Another factor, Western force development policy, can work either of two ways. If NATO does not go ahead with developing and fielding significant numbers of the more advanced conventional weapons, the degradation that the Soviets anticipate in their combined arms doctrine of the 1970s would not occur. If, on the other hand, NATO fields impressive numbers of the weapons, the nuclear issue, heretofore the center of arms control attention, would increasingly be pushed into a secondary place. This trend has been underway for some time. The megatonnage of the U.S. arsenal has been decreasing since the 1960s as the accuracy in delivery systems increased. Now, as the Soviet General Staff sees it, further technological changes could make nuclear weapons unattractive for military purposes.

There is a certain irony in seeing military force developers being more effective than arms control negotiators in reducing the explosive potential of nuclear stockpiles. There is even greater irony in seeing military force developers, through their efforts to make nuclear weapons practical for tactical and operational use, become proponents of more limited and controlled use and perhaps even nonuse of nuclear weapons. The ultimate irony, of course, would be if the West were to make arms control concessions of a kind that would facilitate the modernization of Soviet military forces while denying NATO forces such modernization.

Conclusion

Three major propositions arise from this analysis. Although they might be obvious, they are worth restating.

There is a direct relationship between arms control and trade control. For reasons not altogether clear, there has been a widespread belief among Western analysts that expanded West–East trade would encourage the sort of political change in the U.S.S.R. that would generate Soviet interest in effective arms control and even arms reductions. While there is little historical evidence to support this view, there is massive evidence in both Russian and Soviet history to refute it. Military imperatives have governed much of Russian economic policy, at least since the time of Peter the Great, the tsar who called money the "artery of war." In the Soviet period, as previously indicated, the same has been true even though the ideological rationale was different. Young Soviet Russia in the 1920s feared economic isolation from the West as much as anything else. By concluding the treaties of Rapallo (1922) and Berlin (1926) with Weimar Germany, Moscow avoided a Western coordinated trade policy.[22] Moreover, the Red Army was able to enlist large German credits and technical assistance in the three new military technological areas of the 1920s. The Soviet aviation industry, motor and tank construction, and chemical weapons all depended centrally on German aid. Subsequently, in World War II, the Soviet T–34 tank was considered to be the best.

The infusion of technology from defeated Germany in the aftermath of World War II gave Soviet military industry a needed boost, but that boost had begun to run down by the early 1960s. Within the decade, Western trade and credit expansion were helping to support the new Soviet military modernization. In the 1970s, the expansion became accelerated.

It may be, as many critics insist, that it is impossible to achieve a fully coordinated Western trade policy or even a narrow Western embargo on strategic technology exports. If this is true, then it means that the West can expect both a quantitative and a qualitative arms race for the indefinite future. However, a trade policy that merely slows down the diffusion of technology and credits to the Soviet Union could have a significant effect. The trade control/arms control connection is a reality whether or not this is recognized by proponents of arms control and proponents of extensive East–West trade.[23]

A qualitative arms competition between the U.S. and the U.S.S.R. is likely to make the use of nuclear weapons both less attractive militarily and less probable. The record to date suggests that the qualitative competition between the superpowers is already having this effect. As Samuel Huntington pointed out many years ago, qualitative arms races have tended to be substitutes for war; by comparison, quantitative arms races have tended to lead to war.[24] The point is clear, although it is at odds with contemporary conventional wisdom: some kinds of arms control are not good, even if they are effective in restraining competition. The Strategic Defense Initiative, as well as many of the U.S. Army and the Air Force precision guided munitions and target acquisition systems, would seem to fall into the category of competition that helps to avoid war, not lead to war.

The connection between arms control and arms development is primarily a political rather than a technical matter, a form of political competition. The past decade of arms control experience is a compelling reminder of this proposition. It is difficult to imagine that the present force levels would be higher than they are even if there had been no SALT treaties in the 1970s, and it is possible that without SALT, the Soviet military edge in some categories of forces would not be as great. (It is probable that Soviet ICBMs would be less numerous and the Soviet ABM system technically less advanced.) This need not have occurred. The fault is not so much with arms control as with illusions that an essentially political matter like the East–West military balance can be depoliticized through arms control negotiations. The Western inclination to change the nuclear weapons issue from a political into a technical matter is at the root of the problem. If the West could control this tendency, the chances for successful arms control would be improved. Both the Mutual and Balanced Force Reduction (MBFR) and the Intermediate-Range Nuclear Forces (INF) negotiations are examples of effective Western competition in arms control. Both sets of talks have helped to maintain NATO force levels or redress adverse trends, given Soviet intransigence on arms reductions. In both cases, the West understood Soviet force developments and gave military capabilities first place in designing Western arms control positions. In most other arms talks in the past, the West let technical factors take precedence over military and political realities.

In spite of this, the West has new prospects for success. The Soviet Union has another force modernization task before it, one that will be very costly to execute. New technologies—more than Western military policies—have created this task. Yet, this gives the West a new opportunity to redress the NATO–Warsaw Pact military balance significantly. Changes in U.S. land warfare doctrine are concerned with ways to exploit new kinds of nonnuclear weapons. To make the doctrine effective, NATO has to field a modicum of the new weapons systems and show that it can employ them effectively. At the same time, this opportunity could escape if NATO fails to connect the doctrine rationally to its arms control and trade control policies.

It may appear to some that the West, too, is condemned to the fate of Sisyphus, destined to respond to military build-up after military build-up. In one respect that is true. No single strategy or weapons system will provide security indefinitely. Security is maintained through continuous efforts, frequent reviews, periodic changes in doctrine and strategies, and rhythmic acquisition of weapons and forces using new technologies. Security cannot be bought cheaply, and there are no panaceas to be had, not even in nuclear forces capabilities. There are, however, more effective and less effective approaches to building security for the West. The present juncture, understood from the dynamic perspective of past and potential Soviet force development, seems to offer rare and genuine opportunities for the West to acquire security more effectively. The West can make the Sisyphean task for the Soviet General Staff much heavier and its own task relatively lighter by understanding the rhythm of the competition and by exerting itself in ways that shape the competition. Most important, however, is that the competition be shaped into "a substitute for war" rather than "a prelude to war." The West cannot escape the military competition, but it can compete in ways that make war less likely.

NOTES

1. In a speech printed in *Pravda* (Moscow) on Nov. 30, 1920, Lenin stated: "As long as capitalism and socialism exist, we cannot live in peace; in the end, one or the other will triumph..."

2. Paul Nitze, I was delighted to see, recently discussed this difference in definitions at some length. See his "Living with the Soviets," *Foreign Affairs* (New York), pp. 360-74.

3. The policy of *peaceful* coexistence between socialism and capitalism was proclaimed by Nikita Khrushchev and reiterated by his successors. *Scientific Communism, A Glossary* (published in Moscow in 1975) gives the following official explication of the term: "The CPSU...views peaceful coexistence as a form of class struggle developing in the political, economic, and ideological spheres in the international arena. By fighting against the outbreak of another world war, and organizing and leading the worker, national liberation, and all-democratic movement, the communists...pave the way to the triumph of socialism in the whole world."

4. M.V. Frunze, *Izbrannyye proizvedeniya (Selected Works)*, Moscow, Voyenizdat, 1957, pp. 4-21, 133-42.

5. Trotsky introduced "literacy" classes into the unit training programs and called the Red Army a school for literacy. Officers were encouraged to form "military scientific societies," to publish papers, and to compile their experiences from World War I and the civil war. A general staff college, later to be named the Frunze Academy, was founded for senior staff and command training. See. A. Yovlev, "The Perfecting of Military Educational Institutions in 1921-28," *Voyenno–istoricheskiy zhurnal* (Moscow), No. 2, 1976, pp. 93-98; and Dmitri Fedotoff–White, *The Growth of the Red Army*, Princeton, NJ, Princeton University Press, 1944.

6. For more recent analysis of Soviet military doctrine, see John Erickson,"Soviet Military Operational Research: Methods and Objectives," *Strategic Review* (Washington, DC), No. 5, 1977, pp. 63-73; Peter Vigor, "Soviet Echeloning," *Military Review* (Fort Leavenworth, KS), August 1982, pp. 69-74; Amoretta Hoeber and Joseph Douglass, *Conventional War and Escalation: The Soviet View*, New York, Crane, Russak, 1981; Fritz Ermarth, "The US and the Strategic Balance," a paper presented at a conference on US– Soviet Relations, Washington, DC, March 18, 1983. A paper by Notra Trulock, III, and Daniel Goure, "Soviet Perspectives on Limited Nuclear Warfare," Washington, DC, Nov. 16–17, 1984, is notable for the sources it cites — classified Soviet materials that trace a much earlier Soviet interest in limited and discriminating use of nuclear weapons than is generally appreciated.

7. V.D. Sokolovskiy, *Voyennaya strategiya (Military Strategy)*, Moscow, Voyenizdat, 1962. The volume was published in English translation under the editorship of Harriet Fast Scott with the title *Soviet Military Strategy*, New York, Crane, Russak, 1975.

8. For older references to Soviet war-fighting capabilities, see the citations found in the sources in fn. 6. For quite recent examples, see M.A. Gareyev, *Frunze–voyennyy teoretik (Frunze: Military Theoretician)*, Moscow, Voyenizdat, 1985, and V.G. Reznichenko, Ed., *Taktika (Tactics)*. Gareyev offers a fascinating critique of Sokolovskiy's *Military Strategy* in which he argues that the work was essentially sound in the 1960s but that conditions of war have changed, making many of the traditional principles and concepts of military art, which Sokolovskiy had rejected, once again relevant. In other words, Gareyev is saying that earlier Soviet views on the extent to which nuclear weapons have changed the nature of war and, thus, of military art have proven to be overdrawn; Soviet doctrine is restoring many of the old principles. As my argument proceeds, the centrality of this latest change in doctrine will become apparent.

9. N.V. Ogarkov, *Vsegda v gotovnosti k zashchite otechestva (Always in Readiness for Defense of the Fatherland)*, Moscow, Voyenizdat, 1982.

10. *Ibid.*, pp. 34–35.

11. M.N. Tukhachevskiy, *Izbrannyye priozvedeniya (Selected Works)*, 2 volumes, Moscow, Voyenizdat, 1964.

12. Ogarkov, *op. cit.*, pp. 35, 36.

13. See Stephen M. Meyer, "Soviet Theater Nuclear Forces," *Adelphi Papers*, Nos. 187 and 188, London, The International Institute for Strategic Studies, Winter 1983/84, for a remarkably thorough examination of Soviet capabilities.

14. For Soviet strategic defense and space programs, see *Soviet Military Power*, Washington, DC, U.S. Government Printing Office, 1985, esp. pp. 46-52. For organizational changes in all types of Soviet divisions, see *ibid.* and also W.E. Odom, "Trends in the Balance of Military Power Between East and West," in *The Conduct of*

East–West Relations in the 1980s, Part III, *Adelphi Papers*, No. 191, 1984, esp. pp. 19–20.

15. *Krasnaya Zvezda* (Moscow), May 9, 1984.

16. See Gareyev, *op. cit.* Gareyev specifically mentions choosing the direction of the main effort in the traditional fashion as needing to be revived. Massing and concentrating forces on a main axis, of course, is precisely what echelonment was designed to prevent. Massed forces were too vulnerable to a nuclear strike, in the Sokolovskiy view of war. Insofar as we understand the "operational maneuver group" concept, it amounts to concentrating forces on an axis of the main effort and moving them forward ahead of the dispersed echelons of follow-on forces (see Gareyev, pp. 239–240). Gareyev's critique of Sokolovskiy has a number of other implications that go beyond the argument in this article, but the thrust is the same: pre-nuclear concepts of military art cannot be dismissed in the wholesale fashion dictated by Sokolovskiy's volume.

17. C.W. Donnelly, "The Soviet Operational Maneuver Group: A New Challenge for NATO," *Military Review*, March 1983, pp. 43-60.

18. Gareyev, *op. cit.*, esp. pp. 240–244. Gareyev's revision of Sokolovskiy is an explicit confirmation of my argument about the nature of the most recent changes in Soviet doctrine. A return to more traditional principles of military art — even in the age of nuclear weapons — is, Gareyev says, essential. As a colonel-general, doctor of military science, and professor, Gareyev writes with considerable authority. His book, which was tied in with the commemoration of Frunze's 100th birthday, can be taken as representing the present official line of military doctrine.

19. While neither was more explicit than offering the usual public statements about providing the Soviet Armed Forces with all that they need, the clearest action implying Brezhnev's commitment came with the decision in 1975 to meet economic planning dilemmas by reducing "investment" instead of defense or consumer goods. Andropov did not alter this priority for defense even as he tried to shift resources to free up bottlenecks in energy and transportation. See Myron Rush, "Guns over Growth in Soviet Policy," *International Security* (Cambridge, MA), Winter 1982/83, pp. 167–179.

20. William E. Odom, "Choice and Change in Soviet Politics," *Problems of Communism* (Washington, DC), May–June 1983, pp. 1–21.

21. Criticisms of the Strategic Defense Initiative have been appearing almost daily in the Soviet media since its announcement by President Ronald Reagan. For some examples, see A. Bovin *Izvestiya* (Moscow), June 18, 1985, and the article in *Krasnaya Zvezda*, Mar. 8, 1985, attributed to F. Aleksandrov.

22. Harvey L. Dyck, *Weimar Germany and Soviet Russia 1926–1933*, New York, Columbia University Press, 1966, pp. 50–63. This monograph, based on German Foreign Ministry documents, sounds surprisingly contemporary in the context of East–West trade and arms control.

23. See Office of the Undersecretary of Defense for Policy, "Assessing the Effect of Technology Transfer on U.S./Western Security," Washington, DC, Department of Defense, February 1985.

24. Samuel Huntington, "Arms Races: Prerequisites and Results," in Carl J. Friedrich and Seymour E. Harris, Eds., *Public Policy, Yearbooks*, Vol. 8, Cambridge, MA, Harvard University Press, 1958, pp. 41–80.

10

Soviet and U.S. Military Doctrines and Progress in Arms Control

Vladimir Nazarenko

T he military doctrines of states and their concepts on the development and use of armed forces most clearly reflect their real attitude on the fundamental problems of war and peace. Implementation of new political thinking based on a comprehensive assessment of current realities has required the revision of Soviet military doctrine and the specification of its ideas. Recognizing the fact that war is unacceptable as a means of reaching political ends has compelled us to give up the earlier view of the military doctrine as a system of views on the preparation for and conduct of a possible war.

Modern Soviet military doctrine, as part of the Warsaw Treaty Doctrine, is a system of officially recognized, fundamental views on the prevention of war, military development, preparation of the state and its armed forces to repulse aggression, and on the means of conducting armed struggle in defense of the socialist state. This new definition makes it clear that Soviet military doctrine is aimed at precluding war rather than preparing for it. It is directed at consolidating international security. For the first time in military history, the chief task of the armed forces is to prevent war — both nuclear and conventional. Previously, this task was not so unequivocally determined either in military strategy, in general, or in military doctrine, in particular. Now the aim of our doctrine is to deter an aggressor in order to ensure a sufficient defense.

History shows that any military doctrine is a derivative of policy. The gist and character of Soviet military doctrine is determined by the priority goal of Soviet foreign policy — to ensure for our people an opportunity to work in conditions of stable peace. This fact is reflected in government and party documents and the orders of the supreme military leadership. All practical deeds are strictly geared to them. Let us stress for the sake of justice that, previously, struggle against war was envisaged in the military activities of the Soviet Union. But now that war is tantamount to self-destruction, an all-out catastrophe, the task of preventing a global conflagration has been moved to top priority in our doctrine.

A very important feature of Soviet military doctrine is the defensive orientation of its political and military-technical aspects — the inseparable components of one integral whole.

The political aspect is the main one. It reflects the political goals of the state and defines the character of its effort to enhance national defense proceeding from Lenin's idea that "Any revolution is worth something when it can defend itself." The Soviet leaders believe that no matter how great the threat to peace may be today, a global war is not fatally inevitable — it can be prevented. Political means are taking on significance in solving this task. The military-technical aspect of the doctrine is subordinated to the political one. It includes questions of military development proper, as well as strategic, military-economic, organizational, and technical measures on precluding war and repelling possible aggression. It establishes what armed forces are adequate both for peacetime and for war, specifies ways of preparing and using them, and defines the forms and methods in which they should conduct operations and war in general. A correct answer to these questions is of great practical relevance. Our doctrine is subjected to a thorough analysis in the West not only by military experts, but also by political figures and ideologists. Some of them continue ascribing an offensive and even aggressive character to it. Others recognize the defensive orientation of the Soviet military doctrine, but only regarding its political aspect. The military-technical one is being described by them as offensive.

However, there are no contradictions between the political and military-technical aspects of our doctrine. They form an integral whole and are strictly defensive, both in their goals and ways of solving defensive tasks. Here are the key provisions which make abundantly clear the defensive direction of the Soviet military doctrine.

- First, the Soviet Union does not threaten anyone and will never embark on the road of aggression. It has no territorial claims to any state — either in Europe or beyond. There is not a single state or nation which the Soviet Union regards as its enemy.

- Second, under no circumstances will the Soviet Union be the first to launch hostilities against any state unless it itself becomes a target of aggression. It will never use nuclear weapons first.

- Third, the Soviet Union's military development is strictly geared to the principle of military equilibrium. It is working for military-strategic parity at the lowest possible level for a sufficient defense.

These main provisions of our military doctrine are being implemented in military practice, concrete strategic decisions and actions, plans for the development of armed forces, improvement of their organization and material status, and in operational and combat training. The main requirement is as follows: in solving defensive tasks, the armed forces of the U.S.S.R. proceed from the principle of retaliation, and their development is carried out on the basis of equilibrium and defense sufficiency. The types, quantity and quality of material are strictly tailored to the level of threat and are determined by the requirements for reliable defense and repulse of aggression. Herein lies the quintessence of defense sufficiency. But defense sufficiency must not be interpreted unilaterally, or even less so, understood as a call for one-sided disarmament. Defense sufficiency is closely linked with the principle of military equilibrium.

The entire system of our readiness for defense is being built with a view to preventing an aggressor from reaching its military-political goals regardless of the circumstances under which war is unleashed or conducted. In other words, our combat potential should be permanently maintained at a level ensuring the solution of this task. The main aim of defense is to repel an enemy's onslaught, hold out on the occupied lines, prevent a deep breakthrough, rule out a considerable loss of territory, defeat the attacking groups, roll them back, and restore the state frontier. In this case, the combination of defense and offense does not contradict the defensive orientation of Soviet military doctrine, for the said actions are supposed to be taken against an aggressor.

U.S. Nuclear Strategy

Proceeding from Soviet classification, we may say that the military doctrine of the U.S., which can be described as that of "nuclear deterrence," has obviously preserved continuity in its sociopolitical aspect, that is, definition of the essence, character, and goals of a possible war, in the last 40 years. The most prominent changes were introduced twice: in the late 1950s, when the U.S. recognized that its territory was no longer immune in a nuclear war, and in the late 1960s when the U.S. revised its view on the most probable enemies of the West in Europe and Asia. However, throughout these decades, the notions about the means and methods of reaching ends in war underwent substantial change. This applies to the role of conventional and nuclear arms, development and training of armed forces, and their preparation for basic types of warfare. Strategic, operational, and tactical views (concepts) were changed accordingly. In the 1960s, the strategy of "massive retaliation" was replaced with a strategy of "flexible response." In the first half of the 1970s, the latter gave way to a strategy of "realistic deterrence." In the 1980s, the U.S. administration has not set forth an equitable, clearly formed strategy, although statements by some administration and Pentagon officials have suggested a conventional title for its current strategy as that of "countervailing."

This strategy consists of the concepts of a "limited" and "protracted" nuclear war on a global and regional scale. Under these concepts, war is to be completed on terms advantageous for the United States. With this aim in view, the U.S. is supposed to achieve "a margin of strength in security" (that is, nuclear superiority). Operational concepts of the single, integrated operations plan for nuclear attack and a list of strategic goals envisage different variants of massive and selective counterforce strikes, including preemptive ones. There exist operational concepts of strikes to be dealt by strategic offensive forces against targets of the enemy's conventional potential.

Although U.S. leaders recognize the fact that a nuclear war cannot be won, they still consider its threat to be an instrument of achieving military-political goals. For the time being, the U.S. is not going to give up its prospective military programs, and instead, continues the development and deployment of fundamentally new systems of strategic and tactical weapons. In the first half of the 1980s, the U.S. spent more than one trillion dollars on war preparations. In the current five-year period (1986–1990), it is planning to use almost 1,800,000,000,000 dollars for this purpose. A considerable portion of this sum is designed for upgrading its nuclear potential, especially its so-called strategic offensive triad.

Under the "Comprehensive Program of America's Rearmament" adopted in 1981, the U.S. is developing new systems of first-strike weapons. It is producing B–1B strategic bombers on a massive scale, developing a fundamentally new strategic bomber ATB, deploying M–X ICBMs, and building high-accuracy mobile Midgetman missiles. The Ohio missile submarines are being adopted by the U.S. Navy.

The tactical nuclear systems and general-purpose forces are also being developed. New systems of high-accuracy weapons are being adopted, tactical strike aviation is being upgraded, and the naval forces are being raised to a qualitatively new level.

Space militarization is allotted a place of prominence in the U.S. military plans. On the one hand, the U.S. wants to cover its territory against retaliation with a "space shield," a multilayered ABM system — on the other hand, the U.S. also wants to deploy attack weapons in space, which when combined with strategic offensive arms, would give the U.S. an opportunity to launch a "disarming strike" against the U.S.S.R. and other Warsaw Treaty countries and get away with it.

The offensive orientation of war preparations by and military doctrines of the U.S. and its NATO allies is clearly reflected by their fundamental directives on armed forces development, the latest American and NATO operations and tactical concepts of "airland battle," "air-sea battle," and "advanced maritime frontiers," and other efforts.

The daily activities of U.S. and NATO armed forces are acquiring a dangerous character. It is more and more difficult to distinguish annual exercises in Europe, the Atlantic and other strategically vital regions, from the deployment of troops for warfare.

To sum up, in the 1980s, the military strategy of the U.S. and its NATO allies has become more openly offensive, both in the sphere of strategic nuclear arms and conventional forces. The U.S. and its NATO partners should thus overcome the ideology of "nuclear deterrence" on which their policies have so far rested.

The advocates of nuclear weapons are doing all they can to improve them and draft plans for their use. They specify at what stage of war what types of nuclear weapons should be used, against what targets, and on which scale. In so doing, they want to condition the public into thinking that there is nothing extraordinary about nuclear weapons and that they are essential for security. But in real fact, the concept of "nuclear deterrence" is fallacious, dangerous, and immoral. It expresses the essence of militaristic intentions and embodies a myopic, egoistic approach to the problem of national and international security.

Nuclear deterrence means an unrestrained race for weapons and a chase for military supremacy, a permanent challenge to strategic stability, and perpetuation of international tensions, confrontation, and mistrust.

Nuclear deterrence cannot do without the image of a "potential enemy," the ideology and psychology of antagonism and enmity, the subordination of policy to militarism, and further militarization of thinking.

Nuclear deterrence is urging others to acquire the most destructive weapons, to attain a nuclear status, and to get a potential for threatening others.

Nuclear deterrence leads to unpredictability and uncertainty. It is much like nuclear brinkmanship and only creates a semblance of security and very shaky peace.

Relying on nuclear deterrence means relying on accident, accepting the risk of nuclear disaster which may happen not necessarily due to evil intentions, but as a result of a technical fault. Moreover, technical mishaps will become more and more likely with the further accumulation and sophistication of combat hardware.

Defense Sufficiency and Arms Control

The doctrine of nuclear deterrence has become obsolete. The new political thinking, which rests on the recognition of the necessity and possibility of building a nuclear-free world for mankind's survival, implies resolute renunciation of nuclear deterrence. The new concept proceeds from the fact that the struggle between the two opposite systems is no longer the emphasis of the modern era, that state-to-state relations should be regulated by political and legal mechanisms, international organizations, and other bodies.

In modern conditions, military-strategic parity is a major factor in preventing war. It is primarily expressed in the Soviet–U.S. equilibrium in strategic nuclear forces, which enables each side to inflict unacceptable damage on the other, making pointless any attempt to unleash war. Of course, military-strategic parity as such does not guarantee full security to any state, nor does it rule out the risk of war. In Soviet opinion, the balance of armed forces is impermissibly high, which is fraught with an enormous threat to peace. Apparently, the continuation of the arms race will further enhance this threat and may even bring it to a point where military-strategic parity will cease to be a factor of deterrence.

The Warsaw Treaty countries think that genuine and equal security of states is guaranteed not by the *highest*, but by the *lowest*, level of strategic balance, from which all types of mass annihilation weapons should be excluded.

Today, the only way out is to achieve agreement on drastically lowering the level of military confrontation and reducing military potentials to the limit of defense sufficiency. For strategic nuclear forces, sufficiency implies an ability to prevent the other side from launching a nuclear attack with impunity in any, or even the most unfavorable, situation. Conventional sufficiency means adequate amounts and quality of conventional arms for reliable defense of a country.

No doubt, military confrontation cannot be reduced overnight, but with time, this task can well be solved.

Development of the Soviet armed forces is oriented towards defense. It pays special attention to nationwide air defense troops and groups of forces assigned for defense in different theatres of hostilities, forces and means for antitank defense, engineer troops for building fortifications and entanglements, and motorized infantry formations assigned for defense. To maintain defense sufficiency, it is necessary not so much to supplement the existing structure of the Soviet armed forces with the said elements as to deploy them by means of a corresponding redistribution of troops and weapons.

Meeting the requirements of the military doctrine, we maintain our armed forces in a state of combat readiness which is adequate for preventing a surprise attack. As for radical changes in organization and set-up of our armed forces, they will depend on the level of mutual cuts in troops and weapons which will be achieved and on the real alignment of forces. Here, much will depend on our negotiating partners in the West, above all, in the United States.

The U.S.S.R. and its allies attach primary importance to disarmament, as embodied in the conclusion of a treaty on reducing by half the strategic offensive arms of the U.S.S.R. and the U.S. in conditions of strict observance to the ABM Treaty as signed in 1972 and in nonwithdrawal from it during a negotiated period. The ABM Treaty is a foundation for limiting and reducing strategic offensive arms, without which the development of an ABM system would provoke the build-up of these arms. Nevertheless, the United States seems to be striving to sign the proposed treaty on strategic offensive arms in isolation from the ABM Treaty. The U.S. claims that these arms are not linked to ABM weapons, contrary to the summit agreement whereby the sides pledged to observe the ABM Treaty. By taking part in talks on limiting strategic offensive weapons on earth, the United States is simultaneously trying to justify the continuation of the race for weapons in space. We think that ignoring the link between strategic offensive arms and ABM weapons means deliberately obstructing the elaboration of the agreement made at the talks on nuclear and space weapons.

In addition to this, by displaying reluctance to search for solutions to the problem of sea-based cruise missiles, the proliferation of which destabilizes the strategic situation, the U.S. is impeding success at the Geneva talks. Without the limitation of the Tomahawk nuclear missiles, it is impossible to carry out deep cuts in strategic offensive arms, for in this case, a possibility of circumventing the future treaty would be preserved and a channel for a continued race for strategic weapons would remain opened.

The interests of European and global security require a transition to substantial cuts in troops and weapons in Europe — from the Atlantic to the Urals. It is proposed that the Warsaw Treaty and NATO should drastically reduce their military potentials and retain forces and weapons which are only necessary for defense but inadequate for a surprise attack and the conduct of offensive operations. It is hardly realistic to believe that military alliances could be disbanded today or tomorrow. Mistrust in relations between the U.S.S.R. and the U.S., the Warsaw Treaty and NATO, the East and West is all too great. Needless to say, it will take time to remove the obstacles which were put up in the past. But there is no time to waste. Solution of urgent problems shall not be suspended. Now, the main task is to continue the process of disarmament launched by the INF Treaty, to ensure its continuity and spread it to all types of weapons.

The West often stresses the problem of conventional weapons in Europe, claiming that the Warsaw Treaty has an "overwhelming" superiority in this field. We could, of course, argue against the figures cited by the West on this score, but experience shows that this is useless. Soviet proposals about which our leaders have spoken more than once make it possible to avoid any disputes about figures.

To begin with, it will be expedient to concentrate attention on the mutual elimination of imbalances and asymmetries in individual types of weapons and armed forces of both alliances in Europe. This would create a foundation for tangible mutual cuts in both sides' military potentials.

Then, the armed forces of either side could be reduced by about 500,000 men with organic matériel. Eventually, these troops and weapons should be reduced to a level at which they could only be used for defense.

Measures to reduce and eliminate the risk of a surprise attack are part and parcel of conventional arms reductions. With this aim in view, the U.S.S.R. and other Warsaw Treaty countries suggest creating zones of reduced arms levels, from which the most dangerous destabilizing types of conventional arms would be withdrawn or cut.

A considerable reduction and eventual elimination of tactical nuclear weapons, including dual-purpose systems, are essential for reducing the risk of war and creating a more stable international situation. Finally, the proposals provide for the formation of a system to monitor the observance of agreements with the use of national technical facilities and international procedures, including mandatory on-site inspections.

Although this approach considers the interests of all sides, it has not yet received support from the U.S. and other NATO countries. Apparently, the NATO governments continue to consider conventional weapons as a lawful component of power politics, and keep building them up without fully realizing the consequences of such policy. In the special conditions of Europe, conventional weapons do not differ much from nuclear arms, both in their destructive power and the threat which they pose to the existence of mankind.

One thing is clear: to go from words to deeds in reductions of conventional, strategic, or tactical nuclear weapons, it is necessary to realize that the arms race is both senseless and dangerous and to display the political will to stop and eventually reverse this risky "competition." Such a great military-political success as the INF Treaty gives us the grounds to hope that the international situation will develop exactly in this direction.

11

Rethinking War:
The Soviets and
European Security*

Michael MccGwire

T he Soviet approach to security in Europe is undergoing a truly revolutionary change. Although the Soviets have been explicit about the direction they are trying to go, the fundamental nature and far-reaching implications of this change are not widely appreciated in the West. We fail to recognize its importance because we are so preoccupied with the details of negotiations on strategic and intermediate nuclear force reductions and of the conventional balance in Europe. But we also tend to dismiss the Soviet campaign to promote "new thinking" about international relations as propaganda designed to breed complacency in the West and to give the Soviets the breathing space they need to restructure their economy.

At the center of the new thinking lies the question of national security. The Soviets claim that the traditional approach is outmoded, that a state that unilaterally seeks to improve its security will automatically increase the insecurity of others, and that reliance on military might will actually reduce security by increasing the danger of war. National security, the Soviets now say, can be achieved only by cooperating with other nations to provide mutual security. Guaranteeing national security has "become above all a political task and not a military one." While stressing the need for equal security, the Soviets have moved beyond the idea of parity to talk of sufficiency. They acknowledge the need for asymmetrical reduc-

*This article was reprinted by permission of the Brookings Institution. It appeared in the Spring 1988 issue of the Brookings Review.

tions in arms and for levels that preclude the possibility of surprise attack or general offensive operations. Some quarters are promoting nonoffensive means of defense.

In the world of ideas, none of these is very original. The underlying principles are embodied in the United Nations Charter and have been regularly advocated by Western political leaders, albeit after they were out of office. What is new is that a superpower is proposing these ideas as government policy. More important, they are not some "hare-brained scheme" espoused by a new Soviet leadership, but a set of principles the Soviets can now embrace because of changes in their thinking about the likely nature of a future war. These changes represent the next logical step in the evolution of Soviet military doctrine, a process under way since the mid-1950s.

It is the thesis of this article that the focus of Soviet military policy has moved away from the worst-case contingency of world war toward the lesser contingency of limited war on the Soviet periphery. This shift obviates the need for an offensive posture facing NATO and allows a radically new approach to arms limitations in the European theater.

From 1948 through 1984, planning for Soviet national security centered on the possibility of world war, which would require offensive operations in Europe to defeat NATO. Although the Soviets considered war to be progressively less likely after 1963, the scale of force required to mount a successful offensive against Western Europe steadily increased. A particularly sharp Soviet buildup occurred during the 1970s, a result of a new strategy that relied primarily on conventional means.

In 1983–1984 the Soviets faced up to the fact that in covering the ever more remote contingency of world war, they had raised international tension, fostered distrust, and actually made world war more likely. They had meanwhile incurred heavy political and military costs in the form of NATO's response to what the West perceived as an increase in threat. These political and military costs came on top of the economic burden of maintaining the military capability required to meet their contingency plans for world war.

This logical impasse prompted a shift in military doctrine concerning the possibility of averting world war, allowing the Soviets to recenter their plans on the lesser but more realistic contingency of limited war on the periphery of the Soviet bloc. These plans include the possibility of major conflict on the inter-German border, but the Soviets assume that such conflict will be geographically contained and limited in scope and duration. The invasion of Western Europe, which had been a strategic imperative, has been renounced because it would precipitate world war.

I am not suggesting that the shift in military doctrine is a sufficient explanation for the fundamental changes we are now witnessing in Soviet policies. But for change to occur, the removal of impediments is at least as important as the impulses promoting change. Most of the arguments favoring change have existed for a decade and more, but until recently, they were outweighed by the demands of national security.

Military requirements have generally imposed severe constraints on preferred Soviet domestic and foreign policies, constraints that could be lifted only by changes in military doctrine. This article traces the evolution of that doctrine and the resulting relaxation of specific Soviet military requirements. It will show that General Secretary Gorbachev's concessions in the nuclear arms negotiations reflect the logic of doctrinal decisions reached in the late 1960s, and that the new thinking about security in Europe is not solely the consequence of a generational change in leadership, although that is undoubtedly important.

From the Possibility of World War ...

The West has no equivalent to Soviet military doctrine, which is a continuously evolving system of officially accepted views on basic questions about war. Authorized at the highest political level, military doctrine provides the framework for decisions on strategic and operational concepts, the structure and deployment of forces, and the development and procurement of weapons. At the core of Soviet military doctrine is an assessment of the likely nature of a future war. It is this assessment that determines the way the war will be fought, the range of possible objectives, and the peacetime force posture required to cover the contingency of such a war.

Although Marxist dogma decreed that war between the two social systems was fatalistically inevitable, the Soviets did not think such a conflict was imminent at the end of World War II. Three years later, in 1948, they reversed this sanguine assessment, concluding that by 1953, they would face the real threat of a premeditated Western attack designed to evict them from Eastern Europe, destroy their nascent nuclear capability, and overthrow the Soviet system of government. However, NATO's failure to meet its force goals in the 1950s combined with advances in Soviet military capability to make a deliberate Western attack increasingly implausible. In 1956, the Soviets revised their doctrine to state that world war was no longer inevitable. But they believed that the possibility of world war remained inherent in the antagonism between the two social systems, and should it occur, Western objectives would still include the overthrow of the Soviet state.

World war remained the central contingency for the next 25 years, but Soviet views on its nature and how it should be fought continued to evolve. The initial assessment, which persisted for most of the 1950s, was a projection of the experience of World War II. The basic principles, strategies, and tactics that had proven so successful in driving the Germans out of Russia and back across the eastern half of Europe could be used to repulse a future aggressor and drive him back across the other half of Europe. This thrust into Western Europe would be primarily a byproduct of offensive operations to destroy the enemy's forces rather than an end in itself.

The likely implications of nuclear missile warfare led to a radical reassessment of both the nature of war and the way it would be fought. In the late 1950s, the Soviets concluded that a world war would inevitably be nuclear and involve massive strikes on the Soviet Union. In such circumstances, the optimum Soviet strategy was to launch a preventive nuclear attack if war seemed inescapable. The difficulty of making such a determination and the political constraints on launching such strikes were obvious, but an offensive, damage-limiting strategy was the least of all evils, particularly in the face of their opponents' overwhelming nuclear superiority.

This nuclear strategy did not eliminate the need for a land offensive westward, which had the twin objectives of destroying NATO's military capability and gaining access to Western Europe's economic resources. The latter would be essential to rebuild the devastated economies of the Soviet bloc and to ensure the survival of the socialist system.

Avoiding Nuclear Devastation

Military doctrine underwent another sea change in the late 1960s, when the Soviets decided that a world war would not necessarily be nuclear or involve massive strikes on the Soviet Union. It thus became logically possible, and therefore necessary, for the Soviets to adopt the wartime objective of "avoiding the nuclear devastation of Russia," a corollary being that they would have to forgo nuclear strikes on America because that would invite retaliation against Russia. The U.S. military-industrial base would therefore remain intact, which made it essential that America be denied a bridgehead in Europe from which to mount an offensive once it had built up its forces, as it had in World War II.

The military problem was how to evict U.S. troops from the continent without precipitating intercontinental nuclear escalation. One part of the answer was to restructure Soviet ground and air forces so that ,with conventional means only, they could disrupt NATO's means of nuclear delivery and mount a blitzkrieg offensive into Western Europe. This assault would hamper, perhaps even avert, NATO's resort to nuclear weapons and would reduce the momentum toward escalation. The other part was to deter escalation by threatening retaliation against North America with Soviet intercontinental ballistic missiles. It was this emerging capability, in conjunction with the U.S. adoption of "flexible" and "assured response," that had justified the doctrinal shift.

The need to restructure and increase Soviet ground and air forces to provide the superiority required for a successful conventional offensive meant that the Soviets could not engage in serious negotiations to reduce force levels in Europe in the 1970s. But the very opposite considerations applied to negotiations on strategic nuclear arms. During the 1960s, the Soviets still lagged behind the U.S. intercontinental capability and their ultimate objective was superiority; hence, arms limitations were not in their interests. But the logic of the 1970s strategy argued that Soviet interests would now be served best by nuclear parity at as low a level as possible. Deterrence would work either at a fairly low level or not at all. And if escalation did occur, the smaller the number of U.S. strategic weapons, the less the nuclear devastation of Russia.

The only way to get the United States to reduce its arsenal was to negotiate strategic arms reductions. This military rationale clinched the argument of those Soviets advocating arms control as a means of promoting detente.

Strategic Evidence

The first concrete evidence in strategic weapons policy was the reversal of the Soviet position on antiballistic missiles. The Soviets were concerned that a U.S. ABM system would undermine the credibility of their second-strike capability, and from the earliest days of the strategic arms limitation talks (SALT) in November 1969, they gave priority to reaching an agreement to limit ABM systems. Further evidence appeared in the spring of 1970 when the Soviets abandoned the construction of 78 ICBM silos, indicating they had curtailed their planned deployment of ICBMs targeted on North America, leaving them with a force that matched the U.S. Minutemen in numbers, if not in quality.

The thrust of Soviet strategic weapons policy has persisted from its inception in 1968–1969 through SALT I and SALT II, the strategic arms reduction talks (START) in 1982–1983, the Reykjavik summit October 1986, to the present.

The 78 silos that were abandoned in 1970 included 18 of 24 intended for the "heavy" SS–9, cutting back that program to six years deployment and 288 missiles. The end result was a force of 1,018 third-generation ICBMs targeted on North America, with another 360 adapted to cover regional targets, plus 209 obsolescent, second-generation systems.

In signing SALT II in 1979, the Soviets agreed to reduce their aggregate number of strategic nuclear delivery vehicles (SNDV) to 2,400 (and then to 2,250) and accepted additional cuts in their planned deployment of ICBMs. They had to abort the conversion of 100 silos to launch fourth-generation missiles in order to remain under the newly agreed ceiling for ICBMs carrying multiple, independently targeted reentry vehicles (MIRV). The new sublimit meant that the Soviets had to forgo the replacement of some 550 third-generation missiles.

At the START talks in 1982–83, the Soviets proposed reducing the aggregate number of SNDV to 1,800 and lowering the MIRV sublimit still farther. Their proposal was notable for requiring them to cut their own MIRVed ICBM force from 820 to 680, while accommodating U.S. plans to add 100 M–X ICBMs to its existing force of 550 MIRVed Minutemen IIIs. The current negotiating target is about 1,600 SNDV and 6,000 warheads.

This consistency has been obscured by the need to match the U.S. intercontinental capability in quality (characterized in the West as a buildup). The Soviets were also reluctant to declare their new interest in arms reductions for fear of weakening their negotiating position, while disclosing to the West that they were in the process of adopting a radically new strategy. But throughout these negotiations, the long-term Soviet objective has been parity in offensive weapons at as low a level as possible, together with stringent limitations on ballistic missile defenses.

And the process of arms control has been important in its own right as the only means the Soviets have to effect reductions in the U.S. nuclear arsenal.

Although the Soviets had always pressed for the inclusion of forward-based U.S. systems in the SALT negotiations, their policy on the more general limitation of nuclear weapons in the European theater does not appear to have taken shape much before 1976–1977. By then, Soviet analyses had shown that resorting to nuclear weapons would not help an offensive into Western Europe and would hinder such operations in certain types of terrain. But the primary rationale for limiting such arms remained the danger that resorting to nuclear weapons in the theater would escalate to intercontinental strikes on Russia.

The salience of this issue increased sharply in 1977 when the Soviets started deploying their fourth-generation, intermediate-range ballistic missile, the SS–20, a routine, if belated, replacement for the obsolete first- and second-generation missiles that had been targeted on NATO Europe since the early 1960s. NATO responded sharply, and by 1979, the Soviets were trying to head off a countervailing U.S. deployment by offering to reduce the overall number of missiles facing Europe. Although the complex of calculations that underlay Soviet proposals at the subsequent negotiations on intermediate-range nuclear forces (1981–1983) is obscure, it is relevant that the Soviets were ready to accept significant cuts in existing inventories of nuclear delivery systems.

The implications of the strategy adopted in the 1970s can therefore be characterized as follows. For nuclear weapons, strategic and theater, less was better, and although unlikely to be achievable, zero was best. However, for conventional forces, the opposite applied. If the Soviets were to cripple NATO's nuclear delivery facilities and mount a blitzkrieg into Western Europe using conventional means only, they would need a measure of overall superiority and have to be postured offensively.

Why now are the Soviets talking of military sufficiency in Europe rather than superiority, and why are they proposing to abjure an offensive posture? What has changed?

... to the Possibility of Averting it

In seeking to answer this question, we face the problem that the available evidence is limited to official Soviet pronouncements and the indications of an internal debate. This evidence can be interpreted in various ways, including being dismissed as propaganda. Everyone would prefer to wait for more concrete evidence to emerge, but the rush of events and public pressures do not allow that option. We must therefore postulate the logic that underlies this radical shift in Soviet declaratory policy, drawing on the wider array of world events that are likely to have had some impact on the development of Soviet military thinking.

Central to this approach is the assumption that no leadership, however dynamic, can make radical changes in the higher levels of Soviet military doctrine, except those allowed by developments in objective circumstances and as a part of an evolutionary process. As we have seen, all major changes in doctrine since 1948 have followed this pattern, and we have no indication that these inherent constraints have been removed.

My hypothesis is that the recent developments in Soviet military thinking came about in two stages. The first stage was a byproduct of political-military developments in the early 1980s in the Persian Gulf region, adjacent to the Soviet Union's southern borders. These events raised the question of whether an offensive posture facing Western Europe was still necessary. The second stage, which began in 1983–1984, stemmed from a more fundamental review of international relations. The Soviets made key decisions concerning the political-military level of doctrine in 1984–1985, and debate has now moved down to the military-technical level, which involves practical issues that are much harder to resolve.

Unlocking the Door ...

The first stage was not essential to the emergence of the new military thinking, but it served the important function of "unlocking the door," opening up new avenues of thought that were pursued in the second stage. A backdrop to these developments was the perennial discussion of whether it was realistic even to contemplate fighting a nuclear war.

Before the 1980s objective circumstances had made the possibility of a major regional conflict with the United States extremely remote. While a broad range of possible conflicts had been discussed in the theoretical literature since the late 1960s, it is reasonable to suppose that the practical planning assumption was that major conflict with the United States would involve world war.

In the 1979–82 period, a series of events — including the Iranian revolution, the U.S. embassy hostage crisis, the Soviet invasion of Afghanistan, the formulation of the Carter Doctrine, the institution of a U.S. Rapid Deployment Force, the Iran–Iraq War, the Israeli invasion of Lebanon, and the subsequent involvement of U.S. forces in that area — made a conflict between the two countries in the region north of the Persian Gulf a real possibility. The troubles in Poland added to the general sense of instability, which was exacerbated by the confrontational U.S. posture during this period.

How likely the Soviets thought such a conflict was is not as important as the fact that they were forced to address the issue in practical terms. In these circumstances, one question Moscow absolutely had to decide was whether major conflict with the United States in the Gulf region would necessarily escalate to world war. If escalation was inevitable, then the existing plans for world war still applied and the existing force posture was satisfactory. Once U.S. forces were committed to the Gulf, the Soviets would launch an offensive against NATO. If however, escalation was not inevitable, the Soviets' primary objective would be to contain the war, which would prohibit an offensive into Europe. Instead of moving into Europe, the Soviets would need to hold on their western front, while engaging U.S. forces to the south.

The Soviets concluded that such a regional conflict would not inevitably escalate to world war, implying that they would hold in the west. Although unsurprising, the decision would have acted as a catalyst in two ways. First, it placed in question a tenet of Soviet military thought that had its roots in Tsarist days. Since World War I, a prerequisite for success in a major war was the capability to mount a continental-scale offensive to the west. This enduring requirement underlaid Russian strategy in World War I, was central to Marshal Tukhachevsky's strategic plans in the 1930s, and was triumphantly validated by the experience of World War II. In the postwar period, the ground offensive to the west remained a central concept in the successive reformulations of strategy.

Second, the new requirement to hold in the west raised the question whether forces facing NATO had necessarily to be postured offensively. The military viewpoint would favor the existing posture, since it covered the worst-case contingency of escalation to world war and at the same time pinned down U.S. forces that might otherwise be redeployed to the Persian Gulf. The political viewpoint would emphasize the heavy political and military costs of that offensive posture, costs that were only too evident in the 1979–1982 period.

Since the military-technical factors favored the status quo, there was no operational urgency to resolve that question, but it does seem to have been placed on the agenda. From 1984 onward, a new emphasis on the strategic defensive in ground force operations appeared in unclassified military writing.[1] Since the objective probability of NATO aggression had not increased in the preceding years, the question of adopting a defensive posture is a likely explanation for this development, although more mundane ones are available.

The conclusion that developments in 1979–1982 prompted substantive rethinking is supported by major changes in Soviet command structure during this period. In the early 1980s, Soviet air forces were reorganized in a way that made it easier for the Supreme High Command in Moscow to redeploy the air strike elements between theaters of military action (*teatry voennykh deystviy*, or TVDs), as circumstances might require.

A more far-reaching change was the establishment of three new TVD high commands. Before this period, the Soviets had set up a separate high command only for the Far Eastern TVD — an arrangement that reflected the "world war" concept of holding in the east and fighting in the west. The other TVDs would be run by the Supreme High command in Moscow, as they were in World War II, with the offensive into Western Europe as the focus of concern. But the Soviet decision that regional war would not necessarily escalate to world war introduced a much more complex set of requirements. Although the Soviet plan called for containing a regional war in the Persian Gulf area, operations were likely to spill over into the eastern part of NATO. The Soviets also had to remain prepared to launch an offensive into Western Europe in the event that the conflict could not be contained.

Such a complex strategy would be extremely difficult to run directly out of Moscow. Accordingly, by the fall of 1984, the Soviets had set up high commands for three additional TVDs: the Western TVD, which roughly covers NATO's central front; the Southwestern TVD, covering the Balkans, the western half of Turkey, and most of the Mediterranean; and the Southern TVD, encompassing eastern Turkey, Syria, Israel, and the whole Indo–Arabian region.

The Soviets had been discussing their operational command structure since the second half of the 1970s, so these changes are not conclusive evidence of substantive rethinking. But that explanation would provide a satisfactory answer to various questions: what it was that changed between 1978, when the Far Eastern High Command was established, and 1984; why the boundary between the Southern and Southwestern TVDs runs between Cyprus and the Levant coast, an alignment that is suboptimal for world war but well-suited for the contingency of regional war; and why the Soviets chose not to establish a high command for the fifth continental TVD, covering northwestern Europe and its adjacent seas.

... and Passing Through

The concept of holding in the west effectively downgraded a requirement that had been seen as a strategic imperative for some 70 years. This withdrew a linchpin from the existing framework of geostrategic assumptions and facilitated the second stage in the most recent evolution of Soviet military thinking.

That stage grew out of a review of Soviet foreign policy. Despite the evidence that Soviet policies had been increasingly unsuccessful since at least 1978, the Soviet leadership well into 1983 had remained extremely reluctant to question, let alone jettison, the tenets that had shaped foreign policy through the 1970s. Because of the policy's relative success in the first half of that decade, the leadership had been able to persuade itself that the problem lay in an aberrant deviation of American policy and that "realism" would force the Reagan Administration to moderate its confrontational stance and return to more cooperative policies.

It was probably the Strategic Defense Initiative (SDI) that finally jolted the Soviet leadership out of this complacency.[2] Reagan announced the SDI on March 23, 1983, two weeks after he had described the Soviet bloc as "an evil empire," called its leadership "the focus of all evil in the world," and implied that the Soviet Union was comparable to Nazi Germany. By the end of March, the official Soviet depiction of the danger of war had become sharply more pessimistic, and the very high level of East–West confrontation was explicitly acknowledged at the Central Committee Plenum in June.

The evidence suggests that this plenum sanctioned a formal review of the foreign policy line that had been established at the party congress in 1971 and reaffirmed in 1976 and 1981.[3] This review would have had to acknowledge that developments during 1978–1983 challenged key Soviet assumptions about the structure of international relations. In particular, questions would have been raised about the trend in the correlation of forces, the immutability of détente, and the nature of the superpower relationship.

The review's conclusions can be inferred from the unprecedented "Declaration" General Secretary Yuri Andropov issued at the end of September 1983, a large part of which focused on the threat to world peace posed by U.S. policies. The essence of Andropov's message, however, was that the United States had launched a new crusade against socialism and was bent on military domination. Six weeks later, the Soviet minister of defense reaffirmed this charge, asserting that the United States was preparing for "the elimination of socialism as a sociopolitical system."[4]

This pattern of events argues that Andropov's declaration signaled a decision in principle concerning the continuing validity of the assumptions that had underlain the policies of the 1970s. Past practice suggests that the implications of that decision and the consequent policies would be argued out, and the main implementing decisions made, during the next two years. We have yet to identify, let alone unravel, the full scope of that debate, but there is good evidence that one element concerned world war (see "Reading the Signals").

One aspect of that question was the imminence of war. By 1983, Soviet leaders considered war to be significantly more likely but disagreed on exactly how much more likely. Although important at the time, that aspect need not detain us since what concerns this analysis is the debate on what long-term policies were best suited to *avert the catastrophe of world war*. In seeking to understand the substance of that debate, it is necessary to distinguish between the practical and the theoretical aspects of the dilemma that faced the Soviets.

The Practical Problem

In practical terms, the Soviets found themselves at a logical impasse. A first-order national objective was to avoid world war. Should it be inescapable, the objective was not to lose. By definition, the capitalist objective in such a war would be to overthrow the Soviet system. The Soviets could not, however, respond by seeking to destroy the capitalist system, since that required nuclear strikes on North America, which would invite a retaliatory attack and the nuclear devastation of Russia. The U.S. military-industrial base had to be spared, which implied that NATO must be defeated in Europe and American forces must be evicted from the continent to deny the United States a bridgehead from which to mount a land offensive against Russia. To defeat NATO without NATO's resorting to nuclear weapons, Soviet forces had to be structured for conventional offensive operations and be provided with the necessary superiority.

This threatening posture caused the West to reaffirm its reliance on nuclear weapons, justified U.S. investment in new weapons systems that exploited emerging technologies, and prompted the U.S. concept of "limited nuclear options," which increased the probability of war escalating to an intercontinental exchange. This offensive posture also reinforced the belief that Communists sought military world domination and encouraged the United States to secure its military superiority, this time through weapons in space. East–West tension steadily increased.

The Soviets were caught in a vicious circle. In 1971–1973, they had thought war was most unlikely; 10 years later, they considered it a distinct possibility. In preparing for the remote contingency of a war they could not afford to lose, the Soviets had made world war more likely. Although the Soviets officially acknowledged that there could be no victors in a nuclear war and that it would probably spell the end of human civilization, the possibility of avoiding intercontinental escalation was slim. And even if NATO did not resort to nuclear weapons, the Soviets could not be certain that their offensive in Europe would be successful. If it failed, they would face protracted conventional war against the industrial might of the United States and its allies.

The only way out of this cul-de-sac was to redefine Soviet military requirements. The traditional requirement to avoid defeat in a world war had become an increasingly demanding contingency that determined the size, shape, and posture of the Soviet armed forces. If the military requirement were restated in less ambitious terms, namely, to ensure the territorial integrity and political cohesion of the Soviet bloc, an offensive concept of operations would no longer be needed. That would obviate the requirement for a measure of superiority over opposing forces and would provide new opportunities for confidence-building verification. Secrecy was essential for successful offensive operations, because of the crucial role of surprise; it was much less important in defensive operations.

The Theoretical Problem

The theoretical aspect of the Soviet dilemma involved the ideological doctrine that the danger of war was inherent in capitalism and that the victory of socialism would bring peace to the world. Theoretically, the transition from a world of capitalism to one of socialism, and thereby from latent war to secure peace, was a continuous process that divided into stages, each having a different level of war danger.[5] Achievement of approximate strategic parity in the early 1970s and the move to detente marked the fourth stage, when the capitalists recognized that the outcome of the struggle could not be decided by military means. The next stage required that socialism become the predominant system, and global peace would not be finally assured until the sixth stage, when socialism prevailed throughout the world.

This ideological framework allowed ample room to debate the extent to which peace depended on the achievement of socialism (and vice versa), and to disagree on what went wrong with détente. There was, however, a common acceptance that the fifth stage was not yet imminent, since socialism was still far from being the predominant system in the world. The question was how war could be ruled impossible while capitalism still predominated. The way out of this theoretical impasse was to distinguish between world war and other forms of war, and between the danger of world war and the possibility of neutralizing that danger.

The potential interaction of two new factors allowed the conclusion that the danger of world war could be neutralized. The wide-ranging Western debate in 1979–1983 about nuclear deterrence and the INF deployment had highlighted the growing strength of the peace movements in Europe and North America. But it had also shown that the Soviet Union's conventional superiority, its offensive posture, and its obsessive secrecy had undercut political campaigns to prevent the deployment of Euromissiles. A change in Soviet military policy would not only reduce the likelihood of war by lowering tensions, but could work to neutralize the danger in two ways: by removing an impediment to increased political influence by Western peace movements and by opening up the possibility of a new security regime in Europe and of comprehensive arms control measures.

A Defensive Doctrine

Although some still disagree strongly, the Soviet leadership appears to have accepted the possibility of neutralizing the danger of world war by the fall of 1984 and to have adopted it as an official Soviet objective by the end of 1985. However, the approval of the 27th Party Congress in February–March 1986 was needed to legitimate this new departure, so it was not until April 1986 that Gorbachev was able to unveil the first evidence of a fundamental shift in Soviet policy regarding military requirements in Europe.

In a speech to the East German party congress, he proposed substantial reductions in ground and air forces, accepted the need for dependable verification, including on-site inspection, and acknowledged that the European security problem extended from the Atlantic to the Urals. That definition brought forces stationed in western Russia into the arms control negotiations, while continuing to exclude forces in the United States. In June, the Warsaw Pact proposed that each side reduce forces within Europe by up to 150,000 troops in two years, with a further reduction of 500,000 in the early 1990s, representing a cut of about 25 percent. In this same period, the Soviets changed their policy on intrusive verification, enabling the first stage of the Stockholm Conference on Disarmament in Europe to reach agreement in September 1986.

A further benchmark appeared in May 1987 when the Political Consultative Committee of the Warsaw Pact proposed even more substantial cuts in force levels and defense spending through the year 2000 and recognized the need for asymmetrical reductions. The committee supported Poland's plan for a special disarmament zone between the Rhine and the Bug rivers and issued a formal statement on the Pact's military doctrine, stressing its defensive nature. The Pact's objectives included the reduction of conventional forces in Europe "to the level where neither side, in ensuring its defense, would have the means for a surprise attack on the other side or for starting general offensive operations."

Although the most recent benchmark consists of words, not deeds, and while the claim to a defensive doctrine is hardly new, the significance of the Pact's formal statement should not be underestimated, particularly the objective of denying both sides the capability for surprise attack or a general offensive. If Warsaw Pact forces are to be restructured to support the objective of averting war, the appropriate military doctrine has to be argued out and agreed on at several different levels. The possibility of neutralizing the danger of world war was accepted only in the fall of 1984. When adoption as a formal objective was legitimized in March

1986, the party was already developing military doctrine with a "defensive orientation." And the general principles underlying the new doctrine were accepted less than 15 months later.

This series of decisions mainly affected the sociopolitical level of doctrine, and as the debate moves down to the military-technical level, the room for argument increases sharply. There are professional disagreements on how best to provide for the security of the Soviet Union and jurisdictional disputes about the role of the Communist party in deciding military-technical issues that have traditionally been the preserve of the General Staff. But portraying the present debate as a bureaucratic struggle between the military and the party is too simple. It is better seen as a disagreement on the costs and benefits of military might as the means of ensuring Soviet security, a disagreement that cuts across institutions and generations.

The crucial issue of whether a defensive doctrine necessarily implies the absence of a counteroffensive capability is an important aspect of the argument over what is meant by "sufficiency."[6] But it is noteworthy that, however defined, the concept of sufficiency, rather than parity or superiority, is being advocated as the criterion for sizing forces. It is equally noteworthy that the forces need be sufficient only to provide for the secure defense of the Soviet bloc, rather than a successful offensive into Europe.

The need to escape the political-military impasse has had a major impact on foreign policy, forcing the Soviets to rethink important aspects of their theory of international relations. Marxist theory offers a deterministic explanation of world developments, holding that they are driven by historical forces and broad social movements over which individual nations have no more control than they do over their "natural" behavior. This viewpoint makes it easy to blame the capitalists for all the world's problems, particularly the danger of war, and to argue that Soviet military policies designed to "safeguard the victories of socialism" must necessarily promote the cause of peace.

By adopting the objective of averting world war, even though the preconditions for making war impossible are far from fully developed, the Soviet leadership has implicitly challenged this self-righteous (and inherently passive) approach to foreign policy.[7] One consequence is the new emphasis on *mutual* security and the admission that the Soviet posture facing NATO could be seen as threatening. But it has also legitimized the new thinking about international relations, with its emphasis on global issues and the "interconnectedness" of the world community.

How Should the West Respond?

Two questions now need answering. Has there been a fundamental change in the Soviet approach to national security policy? If so, what should the West do about it?

In answering the first question, we must set aside the evidence of nuclear arms reductions. Although the latest developments are fully compatible with favorable change, they have their roots in the 1967–1968 period and were originally intended to support the 1970s strategy, which called for offensive operations in Europe. Similarly, recent Soviet proposals for mutual reductions of conventional forces in Europe have to be discounted because they would lower the density of NATO's forces, thus improving the chances for a Soviet offensive breakthrough. And, of course, a more conciliatory approach to negotiations promotes progress in both these areas.

There remains, however, substantive evidence of change. The most concrete is the Soviets' radical offer to allow intrusive verification, which they could not have made if they were still wedded to an offensive strategy. Less concrete but still substantive is their willingness to include forces in western Russia in the arms negotiations and their objective, agreed to by the Warsaw Pact, of denying both sides the capability for surprise attack or a general offensive. Finally, we have evidence of the fundamental changes in underlying doctrine and ideological theory that are prerequisites for a major shift in policy. We must add to this the considerable political capital that the Gorbachev leadership is investing in promoting its "new thinking" in international relations.

This body of evidence justifies the provisional assumption that fundamental changes in Soviet policy are now afoot. That these changes represent the logical next step in an evolutionary process is in itself important. Just as gradually lowering the temperature of water will change its state from liquid to solid, so too can an evolutionary change result in a completely new approach to national security.

How should the West respond? Obviously, the Soviet leadership is promoting these policies because it sees them as being in Russia's national interests. The range of potential benefits includes reducing the defense burden, checking the technological arms race, and enabling a new type of relationship with the Western Europeans. Nor have the Soviets put aside the struggle between socialism and capitalism. But because a new security regime in Europe is in Soviet interests does not mean that it is against Western interests. Eschewing a zero-sum approach to East–West relations remains a sound policy, even though it is now being advocated by the Soviet Union.

At this stage the West is not required to do anything very difficult or dangerous. Essentially, we are being asked to sit down with the Warsaw Pact and discuss how to work out a new mutual security regime for Europe involving greatly reduced force levels and major constraints on their structure and posture.[8]

Nobody pretends that such a regime will be easy to achieve. But the discussions should be quite different from the abortive negotiations in the 1970s, because a westward offensive is no longer a Soviet strategic imperative. Success in this enterprise requires that the two sides seek to understand the other's security problems, rather than to focus on unilateral advantage. Much will therefore depend on the nature of the Western response, which will also have an important effect on the internal opposition to the Soviet leadership's initiative. We need to recognize that the Soviets are going through a wrenching reevaluation of their military requirements.

The Soviets have suggested a way out of the military cul-de-sac into which both sides were drawn during the 1948–1953 period. They are espousing means of "nonoffensive defense" that have been developed by nongovernmental circles in the West. Most NATO electorates find these proposals intuitively appealing, and their leaders are clearly being forced to go along, although reluctantly and uncomprehendingly. Change of some kind is now inevitable. If it is to be beneficial, the Western political-military establishment must develop some concept of where it wants to go. It, too, must jettison the thought patterns of the past 40 years and approach the emerging situation with a fresh, yet skeptical, mind. I am not suggesting that we should trust the Soviets — but it is in our interests to suspend *distrust* and to respond imaginatively to their proposals.

NOTES

1. Christopher Donnelly, drawing on his data bank at the Soviet Studies Research Centre, Sandhurst, England. The subject may have been broached earlier in Soviet classified writings.

2. The most important reason Gorbachev advances for new thinking in international relations is that the arms race is on the brink of a qualitatively new and highly dangerous stage, caused by U.S. plans to position "strike weapons" in space.

3. This summary is based on the analysis in Michael MccGwire, *Military Objectives in Soviet Foreign Policy* (Washington, DC, Brookings, 1987), pp. 307-34.

4. D.F. Ustinov, quoted in *Za vysokuyu boevuyu gotovnost* ("For High Combat Readiness"), *Krasnaya zvezda*, November 12, 1983, p. 1.

5. Stephen Shenfield, *The Nuclear Predicament: Explorations in Soviet Ideology* (Routledge & Kegan Paul, 1987), Royal Institute of International Affairs (Chatham House Papers No. 37), p. 27. My summary of the ideological background is based on Shenfield's outstanding analysis of the core values of "peace" and "socialism" in the Soviet ideology of international affairs.

6. This aspect of the debate does divide mainly on institutional lines, with the military advocating "defensive sufficiency" and the defense intellectuals arguing for "reasonable sufficiency." It is important to distinguish between the main argument about security in Europe and the very different question of the strategic nuclear balance.

7. The challenge has now been made explicit in the claim that "class analysis" provides an insufficient explanation of international relations and that the emphasis must be shifted to an analysis of national interests.

8. A.A. Kokoshin, *Sopostavlyaya voennye doktriny* ("Comparing Military Doctrines"), *Pravda,* August 21, 1987, p. 4.

Reading the Signals

Mikhail Gorbachev's election as general secretary in 1985 was followed by overt manifestations of change in Soviet foreign policy, making it logical to link them as cause and effect. A new leadership is not, however, sufficient explanation for the fundamental changes in military doctrine and in the ideology of international affairs that must necessarily underlie this evidence of a new direction in policy. By locating a critical stage of the decisionmaking process in 1983–1984, I am not denying that generational change was a crucial impulse. But generational change began with Yuri Andropov succeeded Leonid Brezhnev in November 1982, and Gorbachev's rise to power is proof that he and his associates were key members of the leadership at this seminal period.

I was predisposed to expect change in 1983–1984 because the analyses documented in my book, *Military Objectives in Soviet Foreign Policy*, had persuaded me that Andropov's September 1983 declaration signaled a decision in principle. Such a decision would be followed by a major reevaluation of Soviet national security policy, leading to a series of "implementing" decisions. Evidence of such a reevaluation can be found in the writings of Marshal N.V. Ogarkov, which, paradoxically, identify both sides of the debate.

Ogarkov had been chief of the General Staff from January 1977 until early September 1984, when he was abruptly reassigned to head the newly formed high command of the Western TVD. During the preceding three years, it had become increasingly clear that Ogarkov held a more alarmist view of the world situation and the necessary Soviet response than did the political leadership, and his reappointment away from Moscow is widely assumed to have contained a disciplinary element. The journal, *Communist of the Armed Forces*, published an article by him in November 1984, the tone and context of which were sufficiently different from his early writing to suggest that Ogarkov had been forced to recant his alarmist opinions. In any case, we have every reason to assume that his statements on the possibility of "eliminating war as a sociopolitical phenomenon" represented official opinion.[1]

The relevant paragraphs of that article were incorporated in a chapter headed "War Can and Must Be Averted," which appeared in a book by Ogarkov, *History Teaches Vigilance*, published six months later. But the wording was changed slightly to warn that the possibility of eliminating war should not be confused with the reality of doing so, and that until that objective is achieved, the danger of war could actually increase. Despite these changes, both statements explicitly state that although the threat of world war remains, it can be neutralized. Ogarkov explains this statement as a "conclusion" drawn by the Communist party about the "absence (*otsutstvie*) of the fatal inevitability of war." In reaching that conclusion, the party had been "guided by the theory of Marxism–Leninism, creatively developing and enriching" the theory "as it applies to the present situation."[2]

Further evidence appeared in *The Second World War: Results and Lessons*, a book with a prestigious editorial board that was written to commemorate the fortieth anniversary of victory in World War II. The book was sent for typesetting in August 1984 and released for printing on February 27, 1985. In the intervening period, the chronology at the back of the book was extended to the end of 1984, and the names of the editorial board were updated to include the new minister of defense and chief of the General Staff. It appears, however, that the six-page "conclusion" was added after the galleys were sent to the printer.[1] It is printed in a different style from the main text and contains a quotation from the special Central Committee plenum that elected Gorbachev on March 11, 1985. Meanwhile, the last numbered chapter of the main text, "Preserving Peace and Averting Nuclear War — The Main Contemporary Problem," has the structure and substance of a concluding chapter.

This added conclusion referred to a "comprehensive analysis of the manifestations and basic tendencies of contemporary social development," on the basis of which the Soviet and other Marxist–Lennist parties concluded that "achieving a durable peace had become a fully realistic task (*zadacha*)." It also notes that "the Communist party is developing military doctrine having a defensive orientation."[3] That use of the present tense and Ogarkov's reference to "the present situation," when combined with the other evidence, persuade me that both Ogarkov and the added conclusion of *The Second World War* are referring to a recent decision and not simply repeating the 1956 doctrine that war was no longer inevitable.[4]

Supporting my assessment of change at this early date was the emergence in 1984 of a new Soviet line about "strategic stability." In the first half of that year, the Soviets also established a new *nomenklatura* of categories of workers in the field of military science and other social, natural, and technical sciences. During this period, the Soviets decreed that "the most important task of Soviet military strategy in contemporary circumstances is to solve the problem of averting war," and more precisely, "averting world war."[5]

NOTES

1. N.V. Ogarkov, "On the 40th Anniversary of the Great Victory; Unfading Glory of Soviet Arms," *Kommunist Vooruzhennykh Sil (KVS)*, no. 21, November 1984, pp. 23-24, 26.

2. Ogarkov, *Istoriya uchit bditel' nosti* (Moscow: Voenizdat, 1985), pp. 85, 90.

3. S.A. Tyushkevic, ed., *Vtoraya mirovaya voyna: itogi i uroki* (Moscow: Voenizdat, 1985), pp. 414, 417. The minister of defense was chairman of the editorial board.

4. The additional evidence includes the implied significance of the "conclusion"; the use of the word *zadacha* (task/objective) rather than the word *tsel'* (goal/objective); and the convoluted and peculiar choice of words regarding the noninevitability of war, a formula that was not used in 1956.

5. Theodore Neeley uncovered the new *nomenklatura* in M.A. Gareev, M.V. *Frunze — voennyy teoretik [M.V. Frunze — Military Theoretician]* (Moscow: Voenizdat, 1985), p. 422. References to "averting war" or "averting world war" are the only substantive amendments to the entries for "military strategy" and "military doctrine" in the 1983 edition of the *Voennyy entsiklopedicheskiy slovar' [Military Encyclopedic Dictionary]*,

2nd ed. (Moscow: Voenizdat, 1986), pp. 240., 712. Raymond Garthoff noted a first-time entry for "strategic stability," p. 703. This concept was being advanced by Soviet defense intellectuals in mid-1984.

Part III

Arms Control Issues
for the 1990s

12

START and Beyond: Implications for U.S. Forces and Strategy

Frank Jenkins

This Administration, as well as each administration in recent times, has cautioned that arms control must never be allowed to become an end in itself. Rather, they have argued, arms control must complement U.S. defense and foreign policy goals. Similarly, since American military forces are intended to carry out our strategy, any limits imposed on those forces — Reagan administration members have stated — must be done in full cognizance of the effect such limits would have on national defense strategy. In other words, we should structure our forces to support national goals and objectives and look for arms control outcomes that can help make the job easier and less costly.

Despite these clear and sound pronouncements, the current administration, as with those in the past, may have lost sight of this basic principle. We now find ourselves having agreed, in two summits, to reduce our strategic forces by fifty percent without any firm understanding of precisely what our force structure should be at those levels. What is worse, we seem to have no real understanding of whether we will be moving toward or away from a more stable strategic balance.

This paper addresses some of the fundamental problems with our arms control approach and provides some suggestions on how to proceed in order to avoid repeating mistakes of the past.

Recent Background

Early in the Reagan Administration, it became clear that arms control proposals should go beyond a mere freeze. Spokesmen for the administration indicated that a freeze was not good enough. Nuclear build down and nuclear freeze proposals were headline grabbers, and the administration wanted to demonstrate that it had more ambitious goals. Reagan spokesmen quickly staked out dramatic positions to prove their point — zero INF and fifty percent cuts in strategic forces. Positions may have been based as much — or more — on political motives than on sound analysis.

In any case, these proposals were adopted and — perhaps due in no small measure to changes in the Soviet Union — the proposals became real candidates for agreement; in the case of INF, they formed the basis for a treaty.

The epitome of how far the Soviets were prepared to go to agree to dramatic arms control outcomes, and how ill-prepared the United States was to judge the outcomes of such agreements, came during the Reykjavik Summit. During these talks, the Soviets offered to eliminate nuclear weapons entirely, and U.S. representatives apparently agreed. The fact that we subsequently withdrew such a notion from the table, or that we might later claim this was nothing more than some cute, high-level jousting, fails to explain either our willingness to move away from a nuclear-based deterrent strategy, what an alternative strategy might be, or the costs of a purely conventional deterrent. We were apparently willing to discuss the most dramatic of arms control outcomes with virtually no analysis to support our proposals.

In any event, we now find ourselves in the somewhat awkward position of having agreed to a START framework calling for fifty percent cuts in strategic forces on the heels of a ratified INF agreement which eliminates the entire class of intermediate-range, land-based missiles, without any real consensus on what our force structure should be under a START outcome. Have we adequately assessed, for example, how our deterrent strategy would be affected? Is the credibility of our extended deterrence weakened? How will outcomes in nuclear negotiations impact the conventional balance and positions to be taken in future Conventional Stability Talks? Given Krasnoyarsk, what is our overall verification policy? I submit that we do not have adequate answers to these questions.

Zero Long-Range INF (LRINF) Outcome

During the late 1970s, NATO sought to deploy a modernized LRINF force, primarily to create a rung that was missing from the escalation ladder. This was the prime argument put forward by Helmut Schmidt in the late 1970s. Arguments concerning the need to offset the Soviet SS–20 force were secondary.

Since our NATO strategy is unchanged, one must presume that U.S. strategic nuclear forces must now be able to fulfill that escalation function in our deterrence strategy. As was the case in 1979, other options, such as the use of sea-launched

cruise missiles, are not credible; they lack the essential ingredient of burden shar-
ing among alliance members that ground-based LRINF provided.

Several questions should be addressed concerning our arms control approach:

- Was the LRINF outcome taken into account when decisions were made
  to propose fifty percent cuts in strategic forces? Did the Joint Chiefs of
  Staff believe that a zero outcome for LRINF was a real possibility when
  they signed on to fifty percent cuts in strategic forces? I question whether
  they did.

- What are the implications of an INF agreement, or any agreement for that
  matter, which lacks real teeth in our ability to verify Soviet compliance?
  What does a very credible cheating scenario mean for the theater balance
  in Europe or the overall U.S.–Soviet nuclear balance? I'm talking here
  about Soviet ability to covertly produce and store new LRINF missiles
  and launchers as they are destroying the old ones.

- What's next for theater nuclear arms control? If zero is appropriate for
  one class of nuclear weapons, why not for all classes? If the escalation control function that had been covered by LRINF can now be fulfilled by
  strategic forces, why is the same not true for shorter-range systems? How
  would the eventual elimination of nuclear weapons from Europe — which
  must be the Soviet goal — affect our ability to assure our allies that U.S.
  security is not becoming decoupled from their security?

Force Structure and Strategy Take a Back Seat

At what point did arms control become an end in itself? At what point did force
structure, strategy, and deterrence take a back seat to our arms control goals?

In addressing these questions, it is important to note that each administration
in the past twenty years has, sooner or later, put arms control high on its agenda.
A key difference is that whereas previous administrations may have viewed arms
control as an end in itself — or actually put arms control in the forefront of their
policy — they put forward or agreed only to limits which had little real military effect. Ironically, it is the Reagan Administration which, despite pronouncements to
the contrary, has established arms control as a political goal while advocating limits
that really would have a significant effect on U.S. forces and strategy.

Deep Cuts in Strategic Forces

I believe the original motivation for deep cuts was primarily political, as I indicated
earlier. Many in the administration saw little or no chance that the Soviets would
accept anything resembling a fifty percent cut.

Here, the administration violated a key rule of arms control: never propose
something you cannot live with — the other side just might accept. The fact that
we do not yet have an agreed-upon U.S. force structure for the levels put forward

in START tells me that we have not done our homework. Either the United States is insincere in its proposal, which I do not now believe, or we are unable or unwilling to decide how best to structure our force if START came to fruition. Mobile ICBMs offer the best example of our inability to decide on key force structure issues. The dilemma created by mobile ICBMs has been recognized and debated for years: mobility can help reduce vulnerability, and thus enhance stability, but mobility can also create verification problems, and hence raise concerns over a different form of instability — namely, rapid breakout.

I must question whether the problem is really that we are uneasy about verification — I believe we simply do not have confidence in our political will to decide on and carry out a mobile deployment in a way that makes sense militarily. I have always believed that some form of multiple aim point system was the best way to satisfy both our strategic and verification goals. I have yet to see any good analysis which refutes that belief.

Implications of Deep Cuts for U.S. Forces

Our inability to arrive at a consensus on a future force structure is a much more serious deficiency today than in the past. There are significant differences between the potential outcomes of a START Treaty and the SALT Agreements, as I've already indicated.

Both SALT I and SALT II actually allowed the United States to carry through with its programs for strategic modernization. There was more than enough "wiggle room" to accommodate changes over the term of the agreements.

With START, major changes will be required just to get down to the limits. More importantly, hard decisions must be made regarding choices among systems, as well as future procurements.

One problem that I see is that we have not done the kind of hard, quantitative analysis required to derive the "best" force structure outcomes (i.e., those that support deterrence, stability, and global strategy). Nor has the United States adopted a long-term competitive strategy which integrates possible countermoves the Soviets might make. As a result, key decisions may be made without benefit of a comprehensive framework for where we want to go.

For example, the 4,900-missile RV number in START will cause us to reduce our nuclear submarine force from the present 36 boats to perhaps 17 or 18. It is not at all clear that the situation will be more stable with, perhaps, 3,500 warheads concentrated on so few targets. Similarly, if the remaining 1,400 or so missile warheads are not deployed on a reasonably survivable ICBM force, stability will be reduced, not enhanced. Yet, in the name of arms control verification, we may either ban mobiles or formulate a scheme that is mobile only in name rather than practice. Mobiles that remain in garrison except during time of crisis can be attractive and enticing targets to an adversary who believes nuclear war might be imminent. Further, while a shift to single-warhead ICBMs would at least be a move in the direction of enhanced stability, there is no broad consensus for such a system. It is simply more cost effective to deploy heavily MIRVed ICBMs in fixed silos, as we have done with M–X. The problem is that what is cost effective will not necessarily enhance stability.

I have used the chart shown below on numerous occasions with a variety of audiences to portray a notional force structure under a START agreement. I show it not so much to argue for this force over any other — but to highlight that it is just as good, or bad, as any force that anyone in the audience might chose to create. It is, in a word, notional. The fact that it is notional highlights U.S. inability to come to grips with key programmatic, R&D, and acquisition decisions that should have been made in advance of, or at least in concert with, our arms control positions.

Table 1. Notional 2005 Force Posture: Impact of START Cuts

System	SNDVs	Weapons per Launcher	Total Weapons	START Rules	START Weapons
ICBMs					
M–X	50	10	500	10	500
SICBM	560	1	560	1	560
Subtotal	610		1,060		1,060
SLBMs					
D–5	480	8	3,840	8	3,840
Subtotal	480		3,840		3,840
Bombers					
B–1B (ALCM)	71	*	1,420	14	994
ATB	100	†	1,600	1	100
Subtotal	171		3,020		1,094
TOTAL	1,261		7,920		5,994

*12 ALCMs/4 GBs/4SRAMs
†8 GBs/8SRAMs

Assumptions:

- Maintain current TRIAD relationships
- U.S. will deploy all 4900 ballistic missile RVs permitted
- ALCM carriers will count under START as carrying 14 warheads but will only count as one SNDV. Penetrators (ATB) will count as one SNDV and one warhead
- Full deployment of mobile SICBM
- 20 Trident SSBNs

Conclusions

The numbers in the START agreement outlined in the Washington and Moscow Summits may or may not be a problem. The point is that we need to understand them better before we sign a treaty.

We need to focus more on U.S. force structure needs and less on trying to solve problems by restructuring Soviet forces. Fifty percent cuts in RVs and cuts in Soviet heavy missiles and throw-weight are good from the standpoint of limiting Soviet breakout. But they cannot ensure the survivability of our own forces — the Soviets would still have more than enough capability to attack our fixed ICBM force.

We need to achieve a broad, enduring consensus now on what our forces should be in the 1990s and beyond — and we need to revisit the START numbers and limits based on that consensus.

Notwithstanding the need for a broad-based consensus we should build into our force structure and arms control planning the flexibility to adopt alternative paths in the event the Soviets violate an agreement or move in directions we had not anticipated.

Key issues are:

- The number of U.S. SSBNs must have some relationship to projected Soviet ASW capability.

- Bomber numbers — and especially bomber loadings — must relate to projections of Soviet air defenses.

- Theater nuclear requirements must relate to extended deterrence, conventional requirements, and strategic forces, all of which, in turn, are tied to allied reassurance.

- We need to clarify the role of the nuclear SLCM. Is it strategic, theater, or both? How many do we really need or want? And, are verification problems truly insurmountable?

- What should our position be on mobile ICBMs? Are we serious about deploying mobiles? What is the right missile? How many? Is our mobile scheme really survivable or simply filling a square? What are the implications of permitting mobiles in a START Treaty and then not carrying through with U.S. deployments?

And, of course, these issues lead to verification questions and concerns. I believe that measures which in the name of verification impact our ability must be secondary to our own force structure and strategic needs. This does not mean that we must concede on key verification measures — simply that we should err on the side of protecting our own force structure requirements, even if the result is less-than-perfect verification.

Finally, I have not addressed in this paper the entire question of strategic and theater defenses, or how the offense-defense relationship should be structured. It is clear that the direction we and the Soviets take on defenses will have a major affect on the strategic balance and the appropriate framework for arms control negotiations and objectives. What is not clear is where the United States will come out on a strategic defense system or the ABM Treaty.

Recommendation

I believe there should be a national debate on our future strategic force structure, taking into account projections of the threat and detailed analyses of how best to carry out our deterrent strategy. The next president should appoint a commission to work with the leaders of Congress and the Department of Defense (DoD) to outline a comprehensive strategic force program for the next 20 years. The effort should take into account a range of Soviet paths and show alternative U.S. approaches and funding needs to cope with changes in the threat and decisions we might take regarding strategic defenses.

One of the concerns frequently raised is how to plan our forces on the one hand while remaining sufficiently flexible to adjust to changing threats. I believe this concern is valid. The problem is that our desire for flexibility is often used to explain away our inability to resolve differences within DoD and between the executive and legislative branches concerning the future of our strategic nuclear forces. I believe we must confront these issues now. We must, in other words, put arms control in its proper place. That is, within a sound force structure framework that is designed to support our strategic and foreign policy goals. It also means, I believe, that our framework for arms control should be sufficiently broad to allow the United States to structure its forces to satisfy its own strategy and doctrine while lowering each side's destructive potential well below today's level. Finally, we must mean it when we say that arms control should not become an end in itself.

13

The Effect of Strategic Force Reductions on Nuclear Strategy*

John D. Steinbruner

T he prospect of cutting strategic nuclear forces roughly in half has seized the attention of the U.S. and Soviet governments, and it is widely accepted that such a step would be, in principle, a good thing. There is much less clarity, however, on the question of exactly why, and that is a warning that the details may prove troublesome. At any rate, the process of deciding which weapons to remove, which to keep, and which must be redesigned will predictably force both political systems to refine their conceptions of strategic security.

The United States in particular will encounter the familiar problem of divided jurisdictions. Strategic weapons are operated by both the Air Force and Navy, which impose significantly different institutional perspectives. The influence of these perspectives can be identified in the details of current reduction proposals. The nominal ceiling that has been agreed upon is 6,000 warheads, but the United States has advanced rules for counting that would allow a substantial number of

---

*This article was previously published in the May 1988 issue of Arms Control Today (Copyright 1988, The Arms Control Association). The numerical assessments referred to in this article are based on an analysis of current and reduced strategic force levels using standard weapons parameters (the yield, accuracy, number, and reliability of warheads and the hardness of targets) and standard methods of calculating strategic force interactions. The detailed assessment was presented in an article by Michael May, George Bing, and John Steinbruner entitled "Strategic Arms Reductions," in Spring 1988 by the Brookings Institute and in condensed version by International Security. The interpretation of the assessment is this article is solely that of Steinbruner.

additional weapons to be carried on bombers, reflecting the Air Force's strong traditional commitment to bomber operations. Moreover, the reductions projected would be roughly proportionate across existing categories and types of weapons, a procedure that makes more sense in preserving the balance of institutional interests than in affecting the calculus of deterrence.

Strategic weapons are purchased in the course of budget cycle, a process that is centered in Washington and is heavily influenced by Congress and all that is involved in domestic American politics. The specific numbers that have emerged have been affected by many purposes — technical investment, industrial employment, and rival fiscal priorities, for example — that are not directly related to the prevention of war. Once purchased, however, the weapons are assigned to targets by a military organization centered in Omaha and held quite remote from public discussion and domestic politics. That organization is professionally driven to do the best it can if ever ordered to fight a nuclear war.

The overall strategic posture of the United States has been influenced by these different organizations and different decision processes, and it has not been perfectly coherent. The familiar doctrine of mutual assured destruction, for example, often proclaimed in the course of debates over budget decisions to be the national strategy, has a distinctly subordinate influence on the process of assigning weapons to targets, as best can be judged from public discussion of that subject.

An agreed reduction of strategic weapons would set direct constraints on the decisions that can be made in the budget process. Congress, in ratifying an agreement, would impose those restraints on itself. That leaves open the question of how the more remote process of weapons targeting would be affected, and hence, what the ultimate strategic consequences would be. That question itself identifies a significant feature of U.S. strategic planning: there appears to be no official process for judging a sufficient level of deterrent capability.

Currently, nearly all the weapons in active inventory are assigned to targets. Those who make the assignments do so by identifying individual targets that have direct military or supporting economic value and by establishing damage criteria for each target. "Requirements" are determined by listing the identified targets, and options are created by provisions allowing categories of weapons to be withheld in the course of carrying out attacks on assigned targets. It is obvious that a lethal effect on the Soviet Union would be achieved long before every individually valuable target was destroyed — just as the death of a living body would occur well before its individual cells were dismembered. The targeting process does not attempt to judge levels of sufficiency for specified missions, however. It sets priorities among basic mission categories, but with that done, the level of target coverage is driven by the number of weapons available and the degree of confidence deemed appropriate in calculating the probable destruction of each individual target. The prospect of strategic force reductions is therefore an occasion for asking an important and officially unanswered question: how many and what types of targets must be threatened in order to achieve the desired deterrent effect?

There is bound to be a spirited division of opinion on this question, at least partially obscured by the veil of classification. With a little common sense, however, it is quite possible to identify the basic arguments in broad outline and to anticipate a plausible outcome. "Plausible" is not the same as "probable," of course, but presumably the one has something to do with the other.

Strategic Missions and Contending Perspectives

There are four basic types of missions that strategic forces can attempt to perform:

- They can attack the strategic forces of the opponent, directly reducing the capability of these forces and thereby reducing as well the damage that might be inflicted on one's own forces and society.

- They can attack the military infrastructure of the opponent — that is, support facilities, conventional forces, and general organizational capability — thereby degrading the opponent's ability to sustain military operations.

- They can attack the industrial capability of the opponent, indirectly reducing military potential but more immediately inflicting punishing damage to people and social organization.

- They can attack specialized targets designed to have some leveraged effect on the opponent while reducing the weight of the attack necessary to accomplish the effect.

Though each of these potential missions can be associated with the idea of deterrence, they differ markedly in the legitimacy and effectiveness that is attributed to them by various segments of American opinion and in the degree of acquiescence that can be expected from the Soviet Union.

The first of these missions (attack on strategic forces), when projected as a strategy for the United States, has the most doubtful legitimacy, the most questionable effectiveness, the fewest domestic advocates, and the greatest Soviet resistance. In the first instance, this mission either directly contradicts or significantly compromises the principle that the United States would act only in retaliation. In each individual instance, an attack on a strategic weapon target must be preemptive to be effective. This strategic concept thus imposes the burden either to initiate war absolutely or to respond so rapidly that it amounts to nearly the same thing. Moreover, because of the protection of strategic weapons installations by mobility and hardening, because of the operational difficulties involved in conducting an attack with the precision and timing necessary to succeed, and because of the strong incentive for the Soviet Union to adopt any feasible countermeasure, the effectiveness of this strategy as a guiding concept for U.S. forces is heavily discounted.

Despite prevailing skepticism about our own ability to carry out a truly decisive attack on Soviet strategic forces, however, there are very prominent fears that the Soviet Union might adopt such a strategy against the United States. This imbalance in judgment is sometimes justified by citing the yield, accuracy, and warhead numbers attributed to the Soviet SS–18 ICBM. The Soviet Union appears to have a corresponding perspective — doubting the effectiveness of the counterstrategic strategy as a guiding concept for their own forces, but fearing it as one that U.S. forces might use. The Soviets cite the M–X ICBM and the prospective Trident II sea-launched ballistic missile as supporting evidence.

These reciprocal fears arise because each side adopts a pessimistic bias in assessing the strategic balance — exaggerating the opponent's potential and discounting their own. In fact, when the total capacities of U.S. and Soviet forces are compared using standard weapon characteristics and standard calculations without a pessimistic bias for either side, the lethality of each force used against the other is similar and well short of that required to eliminate a deterrent threat. With roughly 10,000 strategic weapons currently available, each strategic force could do substantial damage to the opponent's weapons in an initial attack, but particularly because submarines on patrol are not susceptible to swift preemptive surprise, the victim would retain thousands of warheads for retaliation (nearly 3,000 for the United States and more than 1,500 for the Soviet Union). That situation would not be substantially affected by reductions in strategic forces now being proposed. Proportionate reductions to a nominal level of 6,000 warheads would remove available warheads and eliminate opposing targets, but leave the calculated outcome essentially unchanged. That fact facilitates agreement, but reduces the consequences of achieving it. In particular, it means that strategic force reductions of the size and composition now being negotiated are not likely to change the habit of applying a pessimistic bias to strategic assessments and therefore are not likely to conquer reciprocal fears of the counterstrategic mission.

The second of the basic strategic missions — attack on military infrastructure other than strategic weapons — is the primary focus of the theory of deterrence that prevails within the U.S. military establishment. As a broad category, these targets are considered both legitimate and effective. They do not require preemption or a rapid reaction that is nearly equivalent. They can be the focus of strict retaliation. Such retaliation is believed credible and effective in that it would eliminate the organized military capacity of the opponent. That threat is considered by U.S. military officers to be the most powerful and appropriate form of deterrence.

This deterrent threat has not been endorsed by the Soviet leadership, but neither has it been as explicitly rejected as the counterstrategic mission has been. That ambivalence undoubtedly reflects a practical attitude. Any deterrent force of substantial size will unavoidably be very threatening to the coherence of any military organization.

The third basic strategic mission calls for attacks on industrial facilities, and these are also, inevitably, major population concentrations. As a focus for the deterrent threat, this category of targets has dedicated adherents and opponents in the United States. The effectiveness of the attack is not questioned, but both its credibility for purposes of deterrence and its moral legitimacy are sharply disputed. Some believe it is both incredible and immoral to threaten civilian populations. Others argue that such a threat is unavoidable as a practical matter once nuclear weapons are used in significant numbers, and that acknowledging this fact allows effective deterrence to be established at the lowest possible force levels. The latter argument finds some resonance with the general public, but has relatively low standing within the U.S. military establishment.

At the moment, there does appear to be some appreciable difference between the United States and the Soviet Union on this point. At least as can be judged from distant observation, the Soviet political system as a whole is more willing to be guided by a strategic concept that makes the threat of retaliation against industrial targets the principle means of deterrence. Moreover, a practical assessment again must concede that any substantial number of nuclear weapons will threaten the coherence of any industrial society.

The fourth basic type of strategic mission, focusing on targets selected to provide effective leverage at lower levels of attack, plays largely a supplemental role in the United States. In the view of most people, it does not substitute for one of the three other strategic concepts as the main focus for deterrence, but rather it provides buttressing against imagined variations of a typical threat, be it particularly willful opposing decision makers or particularly strained circumstances to which the more general strategic threats may not adequately apply. Opinions on legitimacy and effectiveness are not as well worked out, but some identifiable positions have been established. There are some who believe that petroleum facilities or power installations are appropriate targets for a highly leveraged attack. Others imagine that the deterrent threat might not be sufficiently persuasive unless focused specifically on the opponent's political leadership.

The Soviet Union has been hostile to these arguments for reasons that extend beyond the objective of preserving the deterrent incentive of a strong retaliatory threat. The Soviets are concerned with the possibility that the interactions between the opposing strategic forces could become uncontrollable under crisis conditions. With both sides committed to an extensive process of alerting weapons in response to imminent threat, and both sides contemplating very rapid reaction to any act, the ability to contain the use of nuclear weapons is seriously doubted. For that reason, the Soviet Union resists any attempt to legitimize limited nuclear operations. Because the number of nuclear weapons to be used in such limited options is relatively small, however, this disagreement is unlikely to affect the issue of agreed force reductions.

A Plausible Outcome

Whatever its ultimate fate, the current proposal for reducing strategic forces is not sufficient to settle the underlying issue of relating weapons to targets according to the presumed deterrent effect. Neither the size of the prospective reductions nor the allowed composition of the forces remaining are binding enough to preclude convincingly any of the basic missions. In particular, with 6,000 warheads available to be directed against a reduced number of launchers, and with a quarter or more of these optimized for attacking hard targets, fears of the counterstrategic mission will persist. Political debate will be energized but not resolved; active diplomacy will very likely continue in search of a more decisive outcome.

The outlines of a more decisive outcome are fairly apparent. The obvious objective would be to remove the capacity of each side to conduct an effective attack on the opposing strategic forces, while leaving sufficient capability to accomplish the other possible missions. This would eliminate the most disputed element of the current balance, but would not demand complete resolution of differing opinion about the appropriate focus of deterrence. This outcome could be approximated reasonably well by force reductions to the level of 3,000 warheads, with complete elimination of the advanced multiple-warhead systems (the SS–18, the M–X, and all systems of equal or better combinations of accuracy, yield, and warhead loading) and with rules allowing for mobile basing or, even more probably, for adequately secure basing in silos. Forces of that size and configuration would enable each side to be reasonably confident of their ability to retaliate against 1,500 to 2,000 targets. Attacks of that size directed against military infrastructure targets would so devastate either military organization that it could not expect to conduct coherent operations thereafter. It would inflict damage to civilian populations roughly similar to that produced by a smaller number of weapons targeted directly on industrial facilities and surrounding populations. The competing theories of deterrence would be covered to a degree that would appear to match Mr. Gorbachev's widely noted phrase "reasonable sufficiency."

Such an outcome is unquestionably visionary by the standards of current opinion. Even advocates are likely to consider it more an eventual possibility than an immediate prospect. It contains a logic, however, that, over time, may be a fair match even for the unruly wanderings of American politics and U.S.–Soviet relations.

14

A Soviet View on
Radical Weapons Cuts*

Andrei A. Kokoshin

In the process of reducing nuclear arms, the United States and the Soviet Union will one day reach a stage at which they possess some kind of minimal nuclear forces. And the further we progress in nuclear disarmament, the more important becomes the issue of achieving the most stable configuration of the two sides' strategic forces.

For the sake of strategic stability, such minimal nuclear forces should provide only the capability to deliver a retaliatory strike that would inflict unacceptable damage on the attacker. This means that such forces on both sides must be sufficiently invulnerable and reliable.

The concept for such a configuration of strategic nuclear balance was developed in 1985–1987 by a task force of the Committee of Soviet Scientists for Peace Against Nuclear Threat. It was the result of a study, co-chaired by Roald Sagdeyev, director of the Institute for Space Research of the Soviet Academy of Sciences, and myself, called "Strategic Stability Under the Conditions of Radical Nuclear Arms Reductions."

The study included analysis of a series of combinations of structure and composition in the forces of each side, at several stages of nuclear disarmament. A computer simulation, based on two specially designed models of the strategic balance, was one of the tools used in the analysis.

Reprinted by permission of the Bulletin of the Atomic Scientists, a magazine of science and world affairs. Copyright 1988 by the Educational Foundation for Nuclear Science, 6042 S. Kimark Avenue, Chicago, IL 60637.

The computer simulations applied did not pretend, of course, to be a complete and comprehensive description of the object under examination. Their application was primarily oriented toward determining more precisely the influence of each factor in stability. The use of simulations also helped to identify more clearly the structure of the problem and to determine more tangibly how the effect of a certain stability factor would change, given changes in the parameters.

Taking into account what makes for a stable military-strategic balance, studies of the different stages of nuclear arms reductions should generally adhere to the following criteria:

- The political and military-strategic situation should be such that neither side has any incentive to use nuclear weapons first. The retaliatory actions of the attacked side should preclude the rational use of a first strike.

- Neither side should have the capability to launch a disarming first strike. Whatever the scenario of the attack against one of the sides, the latter must have a potential for inflicting unacceptable and comparable damage to the aggressor.

- There must be no conditions under which the unauthorized or accidental use of nuclear arms could occur.

All three of these criteria presuppose that each side has reliable, redundant, and survivable systems of command, communication, and early warning. The availability of such systems and means, and their purposeful development and perfection on the basis of mutually developed and accepted principles of strategic stability, becomes one of the most crucial factors in insuring stability. In the future, it may be required for this purpose to bring additional means of mutual notification and control into the existing national control systems of both sides.

Assessments of several options for 50-percent reductions in U.S. and Soviet strategic nuclear forces, including those made by means of the computer models, showed that the parameters of the current stable military-strategic balance, achieved by the mid-1980s, would keep their main characteristics in a large number of force combinations. In other words, the attacked side would be able to launch a retaliatory strike, especially if the attack is not totally unexpected, and the retaliator could destroy a wide range of military targets and inflict unacceptable damage on the aggressor's population and industries. It could, in short, commit an act of assured destruction. We presume that for 50-percent cuts, the preservation of the Anti-Ballistic Missile (ABM) Treaty as it was signed in 1972 is an obligatory condition.

It was found, however, that if the side launching a first strike had even a thin territorial ABM system — one that could destroy about half of the missiles launched in a weakened retaliatory strike — the stability of the military-strategic balance would be undermined, unless the other side took certain countermeasures. Among the most vulnerable parts of an ABM system incorporating space-based elements are the command, control, communications, and intelligence subsystems and the battle-management subsystems. The whole ABM system, in general, can be greatly weakened or nullified through the use of a broad spectrum of radio-electronic warfare tactics. Radio-electronic warfare seems to be the best of the countermeasures that are relatively inexpensive and do not call for building an ABM system in response.

The elimination of medium-range nuclear weapons in Europe is an important factor enhancing strategic stability, if strategic nuclear weapons are reduced by 50 percent. This removes from the strategic equation highly accurate systems with a short flight time that threaten the means of early warning and command, control, and communication, which maintain strategic stability by providing the means to assess a situation and prepare a retaliatory strike.

More radical cuts in nuclear forces will have a qualitative effect on the character of the military-strategic balance. For the purpose of analysis, two more hypothetical levels of the military-strategic balance, brought about by further nuclear arms reductions, were also examined: a second 50-percent cut, bringing the strategic arsenals to 25 percent of their current levels, and beyond this second cut, a further reduction bringing the new levels to 5 percent of current strategic arsenals.

If current strategic arsenals are reduced by 75 percent, it is even more crucial that there be no ABM systems — that the 1972 ABM Treaty be strictly adhered to — in order to preserve strategic stability. In addition, taking into account the potential development of antisubmarine-warfare technologies, there will be an urgent need to adopt measures to increase ballistic-missile submarine survivability over what exists now or over what would exist at the stage of 50-percent reduction of strategic offensive weapons. For example, zones should be established in which the antisubmarine activities of both sides are prohibited or substantially limited.

It should be noted that the transformation of the military-strategic balance into a state in which both sides would have approximately as many strategic warheads as they did in the early 1970s would not mean, in the military-strategic sense, a mere return to the situation of that period. The reason for this is that the qualitative characteristics of strategic nuclear forces, primarily in terms of gradually increasing accuracy, are actively changing. There is also a trend toward greater protection of intercontinental ballistic missile (ICBM) silos and of command, control, and communications centers; enhanced abilities to disperse ballistic-missile submarines; and more reliable mobile ICBMs.

According to our analyses, however, higher accuracy is one of the most crucial factors — threatening, above all, the survivability of fixed ICBMs — and it is very difficult to compensate for with other factors. In fact, higher accuracy presents a threat as well to a considerable part of the control and communications systems of the strategic forces of both sides. And accuracy, as it looks now, is extremely difficult, if not impossible, to verify and hence limit by bilateral agreements.

In this connection, at this level of nuclear confrontation, a drastic reduction in the portion of launchers having a large number of warheads with the ability to destroy hard targets early in a conflict is an important condition for insuring strategic stability. This issue, according to calculations and simulations, becomes more critical at the 75-percent stage of nuclear disarmament.

In examining options beyond 75-percent reductions in U.S. and Soviet strategic nuclear forces, the authors of the study proceeded from the premise that from both the political and military points of view, such reductions would be impossible if the other nuclear powers had not by that time participated in the process of nuclear disarmament. Calculations show that in order to maintain a stable strategic balance, it is necessary for the other nuclear powers to reduce their strategic nuclear forces approximately proportionate to the reductions of those of the United States and the Soviet Union, and at the same time to make similar qualitative changes in the structures of their forces.

In the context of such drastic reductions of strategic nuclear forces, it would be unjustified, from the perspective of strategic stability, to maintain the current levels of opposing tactical nuclear weapons systems with a range of up to 1,000 kilometers (620 miles). Although these weapons cannot be used for a disarming nuclear strike against the Soviet Union or the United States, they could be a bridge between a nonnuclear conflict and an all-out nuclear war with the use of ICBMs.

The authors also believe that at this stage of nuclear disarmament, it would become important to take more substantial measures to strengthen strategic stability at the level of general-purpose and conventional forces. The scale of confidence-building measures along the lines of the agreements reached at the 1986 conference in Stockholm should be broadened. In order to lower the probability of surprise attack and offensive operations, there should be reductions and structural changes in the armed forces and conventional weapons of NATO and Warsaw Pact countries. It would also be important to establish additional bans on the development of nonnuclear weapons, based on new or traditional physical principles, whose destructive capabilities bring them closer to nuclear or other weapons of mass destruction.

A 95-percent reduction in both sides' nuclear forces was examined, with the following conditions presupposed: U.S. and Soviet strategic delivery vehicles would only have several hundred remaining nuclear warheads; medium-range and tactical weapons systems of both sides would be eliminated; nuclear weapons of other nuclear powers would be reduced proportionately or completely eliminated, the ABM Treaty would still be in force; there would be a ban on the deployment of space-strike weapons and antisatellite weapons of all types; an agreement on a general and complete nuclear test ban would be in force; and the production of fissionable material for nuclear munitions would be stopped.

At this level, the development of more comparable strategic forces becomes more important, in order to increase the degree of certainty in both sides' assessments of the military-strategic balance. The study's authors concluded, from examining various options, that the best choice for mutual security would be for each side to have approximately 600 light, single-warhead ICBMs, with some portion of these being mobile. All other types of nuclear weapons and their delivery vehicles would be eliminated: bombers carrying gravity bombs, cruise, and ballistic missiles; ballistic-missile submarines armed also with cruise missiles; and other types of nuclear weapons.

Due to the small size and constant movement of light, single-warhead ICBMs, and the peculiarities of the orbits of respective surveillance satellites, it would be difficult to pinpoint the exact location of all these mobile missiles at any given moment. The additional uncertainty stemming from the improbability of a disarming strike would enhance the confidence of each side. And reliable verification of the composition of the strategic forces of each side in such an option, especially when coupled with some additional procedures, is quite possible. The authors did not insist, however, that this is the only possible configuration for minimum-level strategic forces.

ICBMs, especially in fixed silos, are preferable to submarine-launched ballistic missiles in terms of strategic stability, due to the fact that central control over ICBMs is stricter: communications with them are far more reliable and can always be two-way. This means that there is less likelihood of an unauthorized launch, of an accidental nuclear war due to a technical failure or error. In contrast to submarine-launched ballistic missiles, ICBMs would be launched along relatively predictable trajectories. This facilitates the task of early warning for each side, and thus makes the situation more predictable operationally and tactically. In contrast to heavy bombers, ICBMs cannot be engaged in a local, armed, nonnuclear conflict — they are not fraught with the enhanced danger of making a local confrontation global, and a nonnuclear one nuclear.

A strike at ICBMs would be a strike at the enemy's territory and would be the start of an all-out nuclear war. If ICBMs were to become vulnerable, it is technically possible to adopt a launch-under-attack posture — in other words, to launch retaliatory forces while the missiles and warheads of the other side are still in flight. The side planning the first disarming strike could never preclude the possibility of that kind of retaliation.

Another important advantage of this kind of strategic nuclear force structure is that it does not require the inclusion of air defense and antisubmarine-warfare forces and means in the strategic equation, since offensive air and submarine forces would not exist. It would simplify the task of mutually assessing the balance of military power on the basis of common understanding of the strategic criteria and conditions.

The relative weight of air defense and antisubmarine forces increases at low levels of the balance, when strategic nuclear forces have relatively small numbers of heavy bombers and submarine-launched ballistic missiles, even if there is no qualitative improvement in air defense and antisubmarine capabilities.

A mutual shift to light, single-warhead ICBMs would lessen each side's ability to strike the forces of the other, since it would equalize the number of attacking warheads and strategic targets. In this way, the importance of higher warhead accuracy would be substantially diminished. If, however, both sides had more than one warhead on each delivery vehicle and maintained a fixed, ground-based component of strategic forces, the accuracy factor would give them the potential, hypothetical capability of delivering a first strike. Even if the strike may not be totally disarming, it might at least be so destructive that it would reduce the damage of the retaliating attack. Making some part of the ground-based strategic forces mobile during the process of drastic nuclear weapons cutbacks would additionally enhance the survivability of each side's strategic forces if either of them tried to launch a first disarming strike against the other, using warheads of existing technology.

If each side had 500 to 600 warheads on single-warhead delivery vehicles with a high degree of survivability, such a damage-limiting (disarming) first strike would be unattainable. In examining a damage-limiting first strike, one must proceed from the fact that even if such a strike were made, the side delivering it would have no means, as a result of this strike, for dealing follow-up strikes at the industrial targets and large population centers of the other side. At the same time, the attacked side would retain a certain part of its forces capable of launching a retaliatory strike against the main industrial targets and populated centers of the attacker.

The case for deploying a major part of the single-warhead ICBM force in silos is also supported by the fact that third-generation nuclear weapons — nuclear explosions with a powerful electromagnetic pulse — may be capable of short-circuiting the electronic system of mobile missiles. Silo-based ICBMs are easier to screen from such effects than are mobile missiles.

But third-generation nuclear weapons may also endanger command, control, and communications systems. This threat to strategic stability speaks strongly in favor of a comprehensive nuclear test ban, which will block the development of such weapons.

The further build-up, modernization, and diversification of each side's nuclear forces, even while parity is maintained, will lead to a less stable military-strategic and military-political situation, pushing the stability of the strategic balance to the limit. The study by the Committee of Soviet Scientists, however, showed that there are a number of options for radical nuclear arms reductions, up to and including their complete elimination. These would not only permit the balance to stay within the limits of stability, but would broaden those limits as well.

At the same time, their analysis pointed out a number of factors having such a destabilizing effect that the system of military-strategic interaction could lose its stability, given virtually any quantitative or qualitative parameters of nuclear arms. The effect of these factors grows as the process of strategic nuclear arms reduction continues.

One of the most important of these factors is the unilateral or even mutual deployment of territorial ABM systems, especially with space-based components. Because of this, radical reductions are possible only if such destabilizing factors are totally precluded. This is the only way to strengthen strategic stability.

Consistent and complete adherence to the principle of reasonable sufficiency would mean total and universal elimination of nuclear weapons. The point is that each side should have armed forces that are capable only of defending its own territory and that of its allies, rather than those that are capable of threatening other countries with invasion. This goal, however, can only be achieved in a step-by-step process, and at each stage, both sides will have nuclear weapons. At a certain point, the two sides confronting each other would have a minimum nuclear potential capable only of delivering a retaliatory strike and inflicting unacceptable damage on the aggressor. The two sides' nuclear forces guaranteeing a minimum level of deterrence must be sufficiently invulnerable and adequately controlled.

Since verification is key to the process of implementing the principle of sufficiency in nuclear and conventional forces, verification means and procedures should be extended. Verification methods and the modalities of agreements must be such that they prevent any concealed or sudden breach of those agreements, in ways that would change the situation in favor of the violating side.

15

The Military Aspects of European Security After INF*

Robert D. Blackwill

Midway through 1988, a mix of unusual optimism and of some uncertainty pervades East–West security relations. Many of the military slogans and concepts of the past have a faintly antique quality and seem increasingly distant from present circumstances. We should welcome this potential for peaceful change. To remain frozen in the postwar patterns of opposing force structures, weapons systems, and alliances would not only disappoint Western publics, it would also be a futile attempt to defy history's turning wheel.

Nevertheless, like détente in the 1970s, the current condition of peace between NATO and the Warsaw Pact is hardly irreversible; the progressive relaxation of tensions between the alliances is far from inevitable. The Western European psychological and physical landscape can mislead us in this respect. The sound and pain of war in Europe dim in our collective consciousness. The signs of conflict in the cities, villages, and countryside have almost disappeared altogether. Although there have been political altercations along the way, the last East–West crisis in Europe with any military overtones was more than a generation ago, in 1961–1962, over the future of Berlin.

So it is arguable that we in the West have been doing something right in our military policies.[1] But to understand exactly what we have been doing correctly over these decades to turn an endemically warring Europe into a peaceful continent is a controversial endeavor: it is, nonetheless, a principal topic of this essay.

---

*A longer version of this paper was presented at the Annual Conference of the Institute for East–West Security Studies, Potsdam, England June 9–11, 1988.

Whether our motives are benign or bad, if, in either part of Europe, we make destabilizing unilateral changes in our nuclear or conventional military dispositions or implement shaky arms control agreements, we could make this continent and the world more dangerous. Therefore, as we examine ways to enhance military stability in Europe, there is no place for historical ignorance, childish innocence, or utopian prescriptions. Too much is at stake for that — Europe's past is too bloody.

Strategic Nuclear Weapons and Arms Control

Prospects are receding that a START agreement can be concluded and ratified before President Reagan leaves office, although some progress was made at the Reagan–Gorbachev Summit in Moscow. Disagreements between Washington and Moscow continue about the relationship between strategic offense and defense; about mobile missiles; about sea-launched cruise missiles; about counting rules for air-launched cruise missiles; about verification; and about many other lesser issues.[2] At the end of April 1988, 1,200 brackets indicated U.S.–Soviet disputes in the four documents that comprise the draft START treaty.[3] These differences are unlikely to be resolved, and a treaty formulated with sufficient clarity and timeliness to overcome the probable listlessness of the last few months of a two-term American presidency and the effervescent climax of the November U.S. presidential and congressional elections is improbable.

Those who believe such arms control negotiations are too important and technically complicated to be hurried, and who are devoted to the necessary political turbulence of the democratic system, are not inclined to apologize for a hiatus in the START talks.

Still, it is possible to identify, in general, the types of decisions with respect to strategic forces and START outcomes that would promote or weaken East–West military stability. Let us first dispense briskly with the feverish notion discussed at the 1986 Reykjavik Summit that the two superpowers should rid themselves entirely of nuclear weapons, or at least of strategic offensive nuclear weapons. While Mr. Gorbachev's motives can be debated, Mr. Reagan's were simply unfathomable. For, as James Schlesinger has observed in his devastating analysis of the Reykjavik exchanges, "nuclear weapons remain the indispensable ingredient in Western deterrence strategy."[4] This, of course, is not an uncontested assertion everywhere. It should be beyond argument in the White House, and it will be taken as truth in this essay.

Instead, we should devote our attention in START to the question of how the superpowers might significantly reduce the approximately 23,000 strategic nuclear weapons without undermining strategic stability. There are some principles to guide us toward that end:

- A 30- to 50-percent cut in strategic nuclear weapons as envisaged in START I will not increase strategic stability and could weaken it, depending on what modernization takes place in U.S. strategic forces;

- A START I agreement along the lines now being negotiated, though politically important for East–West relations, would neither reduce the threat to ungenerated strategic forces on either side or reduce the likelihood of war over the long term, nor would it save money;

- It is crucial for the United States to modernize its land-based missile force and to make it mobile; this would increase ICBM survivability and enhance the credibility of extended deterrence; and

- Reductions in the next phase of START below the 6,000 accountable weapons limitation of START I could increasingly threaten a secure U.S. second-strike capability, especially if the U.S. does not develop and deploy a mobile ICBM and if there are unforeseen breakthroughs in the technologies of antisubmarine warfare; such strategic nuclear reductions would also place increasing weight on the importance of the conventional balance in Europe and elsewhere.

Beyond INF

The completion of the INF Treaty is the arms control surprise of the postwar period, made possible by Moscow's astounding turnabout. The INF Accord has unquestionably improved the political climate between the United States and the Soviet Union and between East and West. It has also captured massive popular support on both sides of the Atlantic.[5] It marks the first reduction ever of nuclear weapons through negotiation, although in the past two decades, the U.S. has unilaterally removed ten times as many such weapons from Europe as will be eliminated from the American stockpile there by the treaty.

This prospective change in superpower nuclear force structures, although insignificant as a percentage of total arsenals, has induced fear in the elite circles of several NATO nations. Some Western defense experts worry that the INF Treaty squandered the political benefits of INF deployment: it weakened deterrence by eliminating the most credible, European-based U.S. nuclear threat to the Soviet homeland without reducing the Soviet capability to target Western Europe, and it provided aid and comfort to those who wish to push nuclear weapons (and perhaps, eventually, U.S. military forces) off the continent.

That these are, at least to some degree, legitimate concerns does not make them now any less nostalgic. Efforts to increase military stability in Europe must go forward from the reality of the INF Treaty as negotiated, whatever its merit or defects, and not from a rosy remembrance of things past.

Conventional Forces and Arms Control

Interest is rising rapidly on the subject of conventional forces and arms control.[6] As mentioned previously, the INF Treaty has concentrated minds on the conventional balance in Europe. Questions arise again in the West concerning the longevity of the U.S. troop commitment to Europe, particularly in the shadow of

America's huge budget deficits.[7] Without any doubt, however, the most important reason that conventional forces and arms control have vaulted near the top of the East–West security agenda can be found in the person of the General Secretary of the Communist Party of the Soviet Union and his public preoccupation with the domestic Soviet economy.[8] He buried Mutually Balanced Force Reductions (MBFR) in a phrase or two in April 1986, when he accepted General De Gaulle's long-time preference for a European security system that encompassed territory from the Atlantic to the Urals.[9]

Mr. Gorbachev has expounded a wholly new set of Soviet security concepts, including, asymmetrically reducing conventional armaments, minimizing offensive capabilities, eliminating the capacity for surprise attack, structuring solely defensive force postures, and revising military doctrines to give them an entirely defensive orientation.[10] Further, he has so altered general Soviet views on verification that they are no longer recognizable. In short, as no other Soviet leader since Stalin, he has associated himself personally and frequently with the purpose of a radical reduction and restructuring of conventional forces in Europe.[11]

There are good reasons why such a transformation in publicly expressed Soviet concepts first startled, then captivated, much of the West.[12] These ideas from Mr. Gorbachev, if implemented systematically and fairly, would change the political and military face of Europe.[13] They would dramatically cut Soviet conventional forces and armaments, including those massed threateningly in Eastern Europe. They would end decades of Warsaw Pact numerical advantages over NATO in major conventional weapons systems. They would reduce the incentives and the capability for invasion, for surprise attack, for blitzkrieg, and for large-scale offensive action in the early months of a war.

Sad to say, this extraordinary burst of Soviet rhetoric has not yet been matched by positive action or specific proposal. The Warsaw Pact conventional military threat to Western Europe has not diminished during Mr. Gorbachev's stewardship.[14] The most comprehensive and detailed Soviet conventional arms control approach during Mr. Gorbachev's tenure, the Budapest Appeal of June 1986,[15] is strikingly one-sided in design and reeks of "old thinking."[16]

American Secretary of Defense Frank Carlucci apparently encountered the same lack of specificity in his April 1988 meeting with Soviet Defense Minister Dmitri Yazov.[17] This delay in Moscow between idea and action may not be an accurate compass for the future. Relatively little time has passed since the Soviets put these radical concepts forward; implementation necessarily takes longer and may be accelerated after the extraordinary Soviet party conference in late June 1988. Moscow may have been waiting for INF ratification or even a START treaty before proceeding. Mr. Gorbachev and colleagues may wish to engage the new American president on these questions, rather than a president with only a few months left in office.

But the reasons for delay may not be solved by the mere passage of time; they may be more problematical:

- Mr. Gorbachev's difficulties with Mr. Ligachev and those skeptics he seems to represent, Soviet Armenians and other nationalities, the party,[18] the economy, and the defeat in Afghanistan may have weakened his ability to move forward with his reforms concerning conventional forces, if that is indeed his intention;[19]

- There is evidence that the Soviet military is resisting this earth-shaking redefinition of Moscow's European security policy, especially its potential effect on Soviet conventional forces;[20]

- It is unclear how major Soviet withdrawals from Eastern Europe would affect socialist stability there, a fact that must trouble much of the Soviet security elite;[21] and/or

- A cynical Politburo may anticipate less change in Soviet conventional force structure than meets our hopeful Western eyes.[22]

These negative influences on Soviet policy are conjectural and will remain so, absent further development of *glasnost*. However, it is clear that responsible citizens of the West, whatever their innate suspicions, should wish Mr. Gorbachev well in accomplishing what he says in public he wishes to accomplish with respect to equitable reductions of conventional forces in Europe.

The Soviet Union enjoys now, as it has for decades, significant conventional advantages in Europe: in doctrine, numbers of weapons systems, logistics, sustainability, geography, secrecy, and in the likely lack of impediments to its crisis decision making. In addition, most in the West also believe that it is only the Warsaw Pact that would start a war in Europe, and, therefore, the Pact, and not NATO, would choose the time, place, and method of attack and would attempt to achieve surprise. (The notion that the sixteen leaky democracies of NATO could mount a surprise offensive eastward is laughable.)

In addition to those Eastern advantages listed above, many other, less-quantifiable factors, some of which favor the West, would importantly influence the outcome of a conventional war in Europe: technology, training, readiness, mobility, the application of air power to the battlefield, terrain, morale, the quality of the individual soldier as well as that of military leadership, C^3I, alliance integrity, and the unanticipated interaction of all these elements and more — what Clausewitz called "friction."[23] For nonexperts, the mind, like the north German plain, fogs.

Fortunately, the discussion that matters most on this subject is not among Western experts, but between civilian and military officials who represent the members of NATO and of the Warsaw Pact. These East–West exchanges on the elements of the conventional balance in Europe will not reach an agreed conclusion as to the relative conventional strength of NATO and the Warsaw Pact. This subject is too complex and too subjective in character. Both alliances possess perceived strengths and weaknesses they have no interest in admitting to the other, and there is too much self-interest involved. But such discussions, as frank as possible, could help clarify what it is that each alliance genuinely worries most about with respect to the military capabilities of its potential adversary. In this respect, Soviet representatives have made a helpful, if still informal, suggestion: that at the outset of the new conventional arms control negotiations, the two sides exchange data on the forces involved and perhaps even verify those data through an inspection regime.[24]

This detailed approach to information exchange centered on military capabilities is likely to be far more productive as a first step in the talks among the twenty-three participants in MBFR, than it would be to focus on the subject of military doctrine, as the Warsaw Pact indicates in a summit statement it especially wishes to do: "under present-day conditions, it is becoming increasingly important to perceive correctly the objectives and intentions of states and military-political alliances enshrined in their military doctrines."[25] It is not clear why the East believes this would be a fruitful subject to explore, unless, of course, it wishes to use this kind of discussion to undermine with Western publics NATO's current military strategy and, in particular, its reliance on nuclear weapons.

Putting aside that troubling possibility for the moment, it is important to recall that military doctrine reflects in the first instance a nation's (or an alliance's) declaratory policy concerning its military intentions. These public statements of intentions may have little to do with force postures, dispositions, and military capabilities, and still less to do with the private designs and real objectives of the parties to a potential conflict.

Put another way, what needs to change in Europe, as Mr. Gorbachev himself has frequently stressed, is the density and configuration of military forces on the continent, including those at the center. Perhaps a discussion of doctrinal intentions would do no actual harm. But what responsible military officer would recommend a reduction of his country's military preparedness based only on vague public assertions by a potential adversary, including those on military doctrine? Certainly not Soviet Vice Defense Minister General Ivan Tretjak, who warned in February that improved political relations between Washington and Moscow should not cause hurried revisions in the Soviet military such as reductions in active-duty personnel or weakened defenses.[26] We should give these thoughts due consideration.

This is not to say that exchanges on military doctrine have no place at all in East–West negotiations concerning the military aspects of European security. But rather than examining the two alliances' military doctrines in the abstract, such deliberations among the twenty-three nations of NATO and the Warsaw Pact should be inseparable from mutual exchanges of deeply disaggregated data to perhaps the regiment or battalion level (including unit designation, location, subordination, manning figures, and equipment holdings by type and model),[27] and to detailed reductions proposals. Only then would the dimensions and implications of military doctrine acquire the necessary specificity and weight to make this a useful central subject of inquiry and debate between the two sides.

What should be the objectives of conventional arms reductions and limitations in Europe? NATO's stated goals in this respect seem generally consistent with those expressed publicly by the Warsaw Pact. In a March 2, 1988 NATO Summit declaration, Western leaders called for the "establishment of a secure and stable balance of conventional forces at lower levels; the elimination of disparities prejudicial to stability and security; and, as a matter of high priority, the elimination of the capability for launching surprise attack and for initiating large-scale offensive action." [28] If the Warsaw Pact can accept NATO's expressed conventional arms control objectives, the difficult question is how to turn these broadly shared principles into specific proposals and agreements.

The West is alarmed by the massive forward deployment of Soviet divisions and armor in Eastern Europe and by the mobilization potential of Soviet forces in the European portion of the U.S.S.R.[29] Thus, according to the press, discussions within NATO have centered on how to reduce and limit these threatening elements.[30] Such an approach would require very large and starkly asymmetrical cuts in Soviet armor and artillery,[31] perhaps on the order of five to one.[32] James Thomson of the RAND Corporation and the author have proposed equal ceilings on NATO/Warsaw Pact tanks and artillery in the Atlantic-to-the-Urals area and in Central Europe.[33] There are many other schemes around, as well.[34]

Although such reductions in the Warsaw Pact's conventional forces would certainly ameliorate NATO's military problems, Moscow may not find this an altogether attractive idea. The evidence on this score is unclear. *Pravda* has editorialized that "the Warsaw Pact member states are prepared to adhere to the ceilings of sufficiency necessary for defense (and) imbalances on individual types of weapons through reducing the potential of the one who appears to be ahead. As for the Soviet Union, it is prepared for cardinal reductions. The ball is now in the NATO countries' court."[35] Yet, it would be surprising if Mr. Gorbachev and colleagues would simply reduce the Warsaw Pact's armor and artillery to or below current NATO levels without seeking significant compensation from the Western side. It is the precise nature of NATO concessions that is likely to be a major obstacle in the new negotiations among the twenty-three. Outcomes NATO may wish to avoid in the new talks probably do not make a short list. In this phase of conventional arms control, NATO is likely to refuse to accept:

- Major U.S. troop withdrawals from Europe;
- Constraints on future Western European defense cooperation;
- Limitations on force modernization and technological applications;
- The inclusion of aircraft;
- Fundamental structural changes in NATO conventional forces;
- Reductions that would threaten necessary NATO force-to-space ratios, especially on the Central Front;
- Restrictions on NATO, particularly West German mobilization capabilities, on U.S. reinforcement flexibility, and on the transit of U.S. forces through Europe;

- An agreement that could not be verified with confidence;[36] and

- Any link whatsoever between conventional reductions and nuclear weapons in Europe, a subject to which we will return.

Notwithstanding exceedingly positive public statements from Moscow, to imagine the Warsaw Pact agreeing to massive cuts to equal levels with NATO armor and artillery is already a radical notion well outside postwar historical experience. That the Soviets would do so without also insisting that the West give on, say, aircraft numbers and modernization (range, avionics, and payload), naval forces, or nuclear weapons seems close to fantasy. At the same time, NATO concessions in these areas could divide the West, weaken deterrence in Europe and elsewhere, or both. For these reasons and others, a conventional arms control treaty is not on the horizon.

This, of course, does not necessarily mean that there will be no unilateral changes in the conventional force structures of the two alliances. Demography, budgetary constraints, and domestic politics could reduce the size, shape, and location of NATO forces in the period ahead. And there have been intermittent rumors, so far unrealized, that Moscow might pull a few of its divisions out of Eastern Europe,[37] and that Mr. Gorbachev would like to move quickly to reduce the size of the enormous Soviet military establishment. It is sufficient to stress here that the most pertinent question for NATO in the face of any such unilateral Soviet conventional reductions would be whether such drawdowns reduced Soviet military capabilities vis-à-vis Western Europe. The withdrawal of four or fewer Soviet divisions from Eastern Europe would not appreciably lower the threat; neither would major reductions in low-readiness divisions of the Soviet Army far from potential European battlefields or Soviet units deployed along the Sino–Soviet border.

Confidence and Security Building Measures

If negotiating conventional force reductions is going to be a protracted endeavor, some, including the neutral and nonaligned countries, wonder why Confidence and Security Building Measures (CSBMs) garner so little public and political attention in the East or West.[38] Resistance to such an intensified and practical concentration on CSBMs is based on a number of factors:

- The concern over cost – unlike reductions, CSBMs save no money and could, in some cases, be expensive;

- The view that CSBMs would be of little use in a crisis because they would be quickly disregarded by the other side;

- The belief that CSBMs could dangerously constrain military operational flexibility, particularly in a crisis;

- The concern that CSBMs could be overly intrusive and could serve intelligence and even targeting objectives;

- The conviction that CSBMs to provide early warning against an unreinforced attack are not needed since such an attack is increasingly unimaginable, and are of no use with respect to a mobilized attack because there would be innumerable warning indicators of out-of-garrison activities in that situation;

- The worry that CSBMs could be used for purposes of deception; and

- The fact that politicians on both sides of the dividing line in Europe have put CSBMs near the bottom of their European security agendas, preferring to concentrate instead on nuclear, and more recently, conventional reductions.

Broad categories of CSBMs, arguably in ascending order of difficulty to negotiate, include the following: observing more military activities, exchanging information, monitoring permanently selected military installations or locations, and constraining military activities. Long lists of measures in each of these categories have existed for many years, the patrimony in many cases tied to preparations for the 1958 Surprise Attack Conference. There are CSBMs worth considering: exchanging more information on exercises/out-of-garrison activity; more observers and more inspections on demand; placing permanent monitors at militarily important railheads, airfields, equipment storage sites, and ammunition dumps; timely notification of mobilization; and, more problematically, constraints on forward deployment of tanks, artillery, and mine-clearing equipment.[39]

Although it appears unlikely that confidence-building measures in the next phase of the Conference on Disarmament in Europe will markedly change the security landscape in Europe, increased transparency and predictability of the two sides' military forces would promote the cause of stability on the continent. It could also keep East–West discussions on the conventional aspects of European security from quickly relapsing into MBFR's extended paralysis. Though modest objectives, they are worth actively pursuing.

Chemical Weapons in Europe

Because of Soviet concessions under Mr. Gorbachev, progress has been made in the Geneva negotiations on chemical weapons since 1984, when Vice President George Bush tabled a treaty that would eliminate chemical weapons and production facilities globally over a ten-year period. Nevertheless, significant difficulties remain: verification; dealing with agents that could quickly be made lethal; Third World resistance by actual or potential chemical weapons users; and the means to create an international institution that would effectively monitor a chemical weapons ban and treaty.

The verification issue is particularly sticky.[40] To have any hope of guarding against illicit production of chemical weapons would require rapid access of 48 hours or less to virtually any facility on the territory of a possible offender. But to allow such total freedom of movement to inspectors from the other side could compromise sensitive intelligence facilities and operations. If this problem were not formidable enough, there are reputable Western experts who argue that a chemical weapons ban is simply impossible to verify with any confidence. This may suggest that each side should retain a small chemical weapons stockpile for the foreseeable future in case of violations of a chemical warfare accord.

The recent slow pace of the chemical weapons negotiations irritates some Western politicians. Before becoming overly exercised on the subject, these individuals, quite apart from the military implications of undetected cheating, might reflect on the consequences for East–West relations if the U.S. Senate were to refuse to ratify a signed chemical weapons treaty because of doubts about verification. East–West relations hardly need another SALT II experience. Thus, it is best, both militarily and politically, for the West to use caution before it signs a chemical weapons treaty.

Nuclear Weapons in Europe

While conventional and chemical weapons control and CSBMs could face enduring and familiar difficulties, questions relating to nuclear arms in Europe proceed freshly from the INF epiphany. Since no one west of the Urals expected the double zero outcome until Mr. Gorbachev astonishingly accepted NATO's proposal, it is unsurprising that, in the heat of the moment, extravagant claims abound both for and against the INF Treaty. Of more interest now is what happens next concerning nuclear weapons on the continent, especially since there is such a diversity of views between East and West, and within the West, on the subject.[41]

Unlike some mutually shared rhetorical objectives with respect to conventional arms control, the governments of NATO and the Warsaw Pact are vast distances apart on the relationship of nuclear weapons to military stability and deterrence in Europe:

- The Warsaw Pact says it wishes to denuclearize the continent, beginning with the two Germanies;[42] NATO governments argue the virtues of nuclear deterrence and intend to modernize their nuclear forces;[43]

- The Warsaw Pact wants to proceed rapidly to further nuclear negotiations in systems with ranges below 500 kilometers; most of NATO wants major conventional reductions by the East first; and

- The Warsaw Pact may be willing to reduce its numerically superior armor in return for cuts in Western, and particularly U.S., nuclear-capable aircraft; the members of NATO reject this notion out of hand.

That there are disagreements within the West over the nuclear issue obviously offers a temptation to Moscow and its allies. Why not play upon this intra–Western discord in order to weaken NATO? Thus far, this temptation to sow dissension in the West seems stronger than the Soviet Union's will to resist it, notwithstanding the professed Soviet desire to create a less adversarial relationship with the West and to live congenially as a good neighbor within the "common European house."

A concerted Soviet attempt to undermine NATO's military doctrine of flexible response and its essential nuclear component strikes familiar chords of "old" rather than "new" Soviet thinking. How are we to tell the essential difference between Moscow's long 1979–1983 assault on NATO's INF deployment[44] and its current offensive against NATO's other nuclear weapons? Have Soviet long-term military objectives really changed? If they have, why does Moscow concentrate in such a determined way on what it must know is an indispensable element of NATO's political and military integrity? Finally, given Soviet and Czarist history, can the Politburo and senior Soviet military leadership really believe in the reliability of conventional deterrence?

These important questions go to the heart of whether, as all in the West should profoundly hope, Soviet European security objectives have fundamentally changed in these past three years. A continuing drumbeat from Moscow against nuclear deterrence will only strengthen the argument of those in the West more suspicious of long-term Soviet motives and poison the atmosphere of East–West exchanges on future security arrangements from the Atlantic to the Urals. Conversely, a sustained and energetic Warsaw Pact effort to address the most salient and dangerous military instability in Europe, that of Eastern conventional advantages, would do much to demonstrate to Western governments and publics alike that Mr. Gorbachev has indeed revolutionized Soviet European security policy, that he does wish to slash military forces and military competition on the continent dramatically in ways that would lower the risk of war. Moscow's policy on nuclear weapons in Europe and its willingness to accept large and deeply asymmetrical cuts in its conventional forces, including those in Eastern Europe, will be the most accurate litmus tests to identify actual Soviet security objectives in Europe in the years ahead.

Prospects

An American baseball figure once observed that "predictions are dangerous, especially about the future." So, instead of trying to guess the direction that Mr. Gorbachev, the next U.S. president or the nations of NATO, the Warsaw Pact and the Conference on Security and Cooperation in Europe (CSCE) will take European security, suffice it here to list some principles that might guide what should be a thorough and aggressive search by all for increased East–West military stability in Europe:

- With respect to military forces in Europe, avoid attaching importance to declaration of intent, vague political pronouncements or quests for a distant Elysium;

- Stick instead to the examination of concrete and practical measures — the more detailed, the better;

- Concentrate on reducing significantly the asymmetries in conventional forces and increasing their transparency and predictability;

- Expect force modernization to continue;

- Maintain credible nuclear forces on the continent and nuclear deterrence; and

- Above all, do not, through arms control, unwise changes in force posture, or attempts to achieve one-sided advantage, make a stable European continent unstable.

NOTES

1. A brilliant book on the general postwar evolution of East–West relations in Europe is A.W. Deporte, *Europe Between the Superpowers: The Enduring Balance* (New Haven: Yale University Press, 1979).

2. Henry A. Kissinger's critical view of the START I process is enumerated in "Policy Must Precede New Arms Treaty," *Los Angeles Times*, April 24, 1988, Part 5, p. 1.

3. R. Jeffrey Smith, "As Schultz Heads for Moscow, Reagan Is Said to Stand Firm on Arms Proposals," *Washington Post*, April 19, 1988, p. A-8; and Fred Kaplan, "US Hope Fades for Arms Pact Before Summit," *Boston Globe*, April 23, 1988, pp. 1, 6.

4. James Schlesinger, "Reykjavik and Revelations: A Turn of the Tide?" *Foreign Affairs*, Vol. 65, No. 3 (1987), p. 429.

5. The best scholarly justification for the INF Treaty is Graham Allison and Albert Carnesale, "Can the West Accept *Da* for an Answer?" *Daedalus*, Vol. 116, No. 3 (Summer 1987), pp. 69-93. For a critical evaluation of the INF agreement, see Bernard W. Rogers, "Arms Control and NATO Deterrence," *Global Affairs*, Vol. III, No. 1 (Winter 1988). A Soviet perspective on the Western debate concerning the INF Treaty is in Nikolai Portugalov, "Double Strategy — Double Error," *New Times*, Vol. 43, No. 87, pp. 12-15.

6. For reasons for this, see John Borawski, "Farewell to MBFR?" *Arms Control Today*, May 1987, pp. 17-19.

7. "Europe's Brave Colours," *The Economist*, July 11–17, 1987, pp. 11-12.

8. In a February 1987 speech, Gorbachev noted that "our international policy is more than ever determined by our domestic policy, by our interest in concentrating on constructive endeavors to improve our country ... This is where we want to direct our resources, this is where our thoughts are going, on this we intend to spend the intellectual energy of our society." Philip Taubman, "Gorbachev Avows a Need for Peace to Pursue Reform," *New York Times*, February 7, 1987, p. 1. Many of the same themes are also in Mikhail Gorbachev, "Address to Media Chiefs' Meeting," *Pravda*, January 13, 1988, pp. 1-3; and in Gorbachev's "Address to Plenum on Restructuring," *Pravda*, February 19, 1988, pp. 1-3.

9. Jackson Diehl, "Gorbachev Says U.S. Acts Hurt Ties But Offers a 'New Initiative' on Forces," *Washington Post*, April 19, 1986, p. A-21.

10. For optimistic views concerning change in Soviet policies on conventional forces and arms control, see Jack Snyder, "Limiting Offensive Conventional Forces: Soviet Proposals and Western Options," *International Security*, Vol. 12, No. 4 (Spring 1988), pp. 48-63; Jerry F. Hough, "Soviet Arms Control Policy in Perspective," *The Brookings Review* (Winter 1988), pp. 39-45; and Michael MccGwire, "Rethinking War: The Soviets and European Security," *The Brookings Review* (Spring 1988). More skeptical

treatments of Moscow's objectives are in David S. Yost, "Beyond MBFR: The Atlantic to the Urals Gambit," *Orbis* (Spring 1987), pp. 99-134; Karl–Heinz Kamp, "Prospectives of Conventional Arms Control in Europe," *Assenpolitik* (English Edition) Vol. 4 (1987), pp. 331-343; and Lev Yudovich, "Warsaw Pact's New Military Doctrine: More Velvet Glove, Less Iron Fist," *Armed Forces Journal*, February 1988, pp. 380-39.

11. Dale R. Herspring argues that *perestroika* could increase the Soviet military threat to the West in his article, "On *Perestroika*: Gorbachev, Yazov, and the Military," *Problems of Communism*, July–August 1987, pp. 99-107.

12. A particularly breathless and naive U.S. reaction to Gorbachev's rhetoric appears in H.D.S. Greenway, "Gorbachev's New Military Doctrine," *Boston Globe*, December 4, 1987, p. 19. A somewhat more measured press treatment is Don Oberdorfer, "Soviets Signal Major Shift in Military Policy," *Washington Post*, November 30, 1987, pp. A-1, A-4.

13. The broader Soviet international approach, in the context of which conventional forces would be restructured, is explained in Y. Primakov, "New Philosophy of Foreign Policy," *Pravda*, July 9, 1987, p. 4. A discussion of the potential political effects in Europe of conventional arms control is in Robert D. Blackwill, "Conceptual Problems of Conventional Arms Control," *International Security*, Vol. 12, No. 4 (Spring 1988), pp. 29-34.

14. This point is made by SACEUR General John Galvin in Serge Schmemann, "NATO Chief Is Wary, Treaty or No," *New York Times*, April 28, 1988, p. A-7.

15. Excerpts of the Budapest Appeal are published as "Address of Warsaw Treaty Member States," *Survival*, Vol. 29, No. 5 (September/October, 1987), p. 463.

16. For an analysis of the Budapest Appeal, see Robert D. Blackwill, "European Security and Conventional Arms Control," James A. Thomson and Uwe Nerlich, eds., *Conventional Arms and the Security of Europe* (Boulder, CO: Westview Press, forthcoming).

17. "Superpower Military Talks," *New York Times*, April 20, 1988, p. 28.

18. Bill Keller, "Gorbachev Would Ease Grip of Party Apparatus," *New York Times*, April 28, 1988, p. 6.

19. A typical description of Gorbachev's problems is in Philip Taubman, "Gorbachev Effort to Reshape Soviet Facing Paralysis," *New York Times*, April 17, 1988, p. 1; in Taubman, "No. 2 Man in Soviet Cuts His Activities," *New York Times*, April 21, 1988, p. A-13; and in Taubman, "Soviet Politburo Is Said to Demote the No. 2 Leader," *New York Times*, April 22, 1988, pp. A-1, A-8. A more thoughtful and rigorous analysis of the subject is Seweryn Bialer, "Gorbachev's Move," *Foreign Policy*, Number 63 (Fall 1987), pp. 59-87; and Bialer, "Inside *Glasnost*," *Atlantic Monthly*, February 1988, pp. 65-74. See also John B. Dunlop and Henry S. Rowen, "Gorbachev Versus Ligachev," *The National Interest*, Spring 1988, pp. 18-29; and a letter by V. Selivanov in *Pravda*, May 2, 1988, p. 1, which, in a remarkable attack on the Soviet party system, asserts that "the truth is that our errors and failures are, in the first instance, the failings of our party and the Central Committee and each party member."

20. Robert Legvold, "Gorbachev's New Approach to Conventional Arms Control," *The Harriman Institute Forum,* Vol. 1, No. 1 (January 1988), pp. 1-8; and several personal conversations with Soviet representatives.

21. There are, of course, plenty of other problems already in Eastern Europe; John Tagliabue, "Eruption of Labor Unrest Leaves Warsaw Unsettled," *New York Times,* April 28, 1988, pp. A-1, A-7.

22. Jim Courter, "The Gathering Storm: Are the Soviets Preparing for World War III?" *Policy Review,* Fall 1987, pp. 1-9.

23. Carl von Clausewitz, *On War,* rev. ed.,. Michael Howard and Peter Paret, eds. and trans. (Princeton, NJ: Princeton University Press, 1984), pp. 119-121.

24. Bulgarian Deputy Foreign Minister Ivan Ganev in late March 1988 promised Warsaw Pact data "very soon," as quoted in "East Bloc Vows Data on Forces," *Chicago Sun Times,* March 31, 1988, p. 36. Also based on personal conversations with Soviet representatives in Vienna and elsewhere.

25. "Session of the Political Consultative Committee of the Warsaw Treaty States," *Documents* (Berlin: 28–29 May 1987), pp. 11-14. For a military elaboration of that summit document, see Sergey Fedorovich Akhromeyev, "The Doctrine of Preventing War, Defending Peace and Socialism," *Problemy Mira I Sotsializma,* Vol. 12 (Moscow: December 1987), pp. 23-28; and Colonel General Dmitri Volkognov, "View from the Other Side: The Warsaw Treaty's Antiwar Doctrine," *Armed Forces Journal,* December 1987, pp. 48, 50. An excellent historical treatment of the development of Soviet military doctrine is Condoleezza Rice's "The Making of Soviet Strategy," in Peter Paret, ed., *Makers of Modern Strategy* (Princeton, NJ: Princeton University Press, 1986), pp. 648-676.

26. Tretjak interview with *Moskovskie Novosti* as reported in *La Stampa,* February 18, 1988.

27. The exchange of large lumps of aggregated data would have little utility, and one hopes this is not what the Warsaw Pact has in mind.

28. "Conventional Arms Control: The Way Ahead," Statement Issued Under the Authority of the Heads of State and Governments Participating in the Meeting of the North Atlantic Council in Brussels, March 2–3, 1988, pp. 4-5.

29. One of the many articles on the subject is Peter A. Peterson and John G. Hines, "The Conventional Offensive in Soviet Theater Strategy," *Orbis,* Vol. 27, No. 3 (Fall 1983), pp. 695-739.

30. For instance, William Tuohy, "NATO Aiming for Conventional Arms Plan that Soviets Could Accept," *Los Angeles Times,* January 13, 1988, pp. 10, 14; and Peter Adams, "NATO Will Focus on Equipment Ensuring Against Surprise Attack," *Army Times,* February 1, 1988, p. 29.

31. *Ibid.*

32. James A. Thomson and Nanette C. Gantz explain the military rationale for this ratio in "Conventional Arms Control Revisited: Objectives in the New Phase," A RAND Note N-2697-AF (Santa Monica, CA: The RAND Corporation, December 1987). See also Michael R. Gordon, "Study Says Troop Pact Is Not Likely," *New York Times,* November 12, 1987, p. A-3.

33. For the details of our approach, see Robert D. Blackwill and James A. Thomson, "A Countdown for Conventional Arms Control," *Los Angeles Times*, October 25, 1987, Outlook Section, pp. 1, 2, 6; my testimony on the INF Treaty and that of Dr. Thomson before the Senate Armed Services Committee, February 17–18, 1988; and "Across the Great Divide: NATO on the Imbalance of Conventional Forces," *Newsweek*, March 14, 1988, p. 33.

34. For instance, Jonathan Dean, "Will Negotiated Force Reductions Build Down the NATO–Warsaw Pact Confrontation?" Presentation for the House Subcommittee on Conventional Forces and Alliance Defense, November 3, 1987; Karl Kaiser, "Objectives, Concepts, and Policies for Conventional Arms Reductions," Testimony Before the Senate Armed Services Committee on the Occasion of its Hearings on NATO Defense and the INF Treaty, February 17, 1988; Albrecht A.C. von Muller, "Conventional Stability in Europe, Outlines of the Military Hardware for a Second Détente," Institute Paper, Starnberg, 1987; Stanley R. Sloan, "Conventional Arms Control and Military Stability in Europe," Congressional Research Service, October 16, 1987; and so forth.

35. As quoted in Matthew C. Vita, "Soviets Show Flexibility on Cutting Conventional Forces," *Atlanta Journal and Constitution*, December 20, 1987, p. 8. Mr. Gorbachev himself stresses that "if an imbalance and disproportions exist, let us remove them." This quote is from an important Gorbachev statement on the application of new thinking on international relations, M.S. Gorbachev, "The Reality and Guarantees of a Secure World," *Pravda*, September 17, 1987, pp. 1-2.

36. The great difficulty in verifying a conventional arms control agreement is discussed in Robert D. Blackwill, "Conceptual Problems of Conventional Arms Control," *International Security*, Vol. 12, No. 4 (Spring 1988), pp. 43-47.

37. Michael T. Kaufman, "Gorbachev on a Prague Stroll: The Cheers and the Curiosity," *New York Times*, April 10, 1987, p. A-8.

38. A comprehensive assessment of the role of CSBMs in European security is by James MacIntosh, "Confidence (and Security) Building Measures in the Arms Control Process: A Canadian Perspective," prepared for The Arms Control and Disarmament Division, Department of External Affairs, Ottawa,Canada, August 1985. See also two recent Institute for East–West Security Studies (IEWSS) reports, Rolf Berg and Adam-Daniel Rotfeld, "Building Security in Europe, Confidence-Building Measures and the CSCE," ed. Allen Lynch, IEWSS, New York, 1986; and F. Steve Larrabee and Allen Lynch, "Confidence-Building Measures and U.S.–Soviet Relations," IEWSS, New York, 1986. A Soviet perspective on confidence building is Vladimir Petrovskiy, "Confidence and the Survival of Mankind," *Mirovaya Ekonomika I Mezhdunarodnoye Otnosheniya*, Vol. 11 (November 1987), pp. 15-26. Like most Soviet statements on the subject, it is exceedingly general in character.

39. A menu for possible future CSBMs can be found in John Borawski, "Practical Steps for Building Confidence in Europe," *Arms Control Today*, March 1988, pp. 17-18. Borawski has written widely on this subject.

40. See Charles C. Floweree, "Elimination of Chemical Weapons: Is Agreement in Sight?" *Arms Control Today*, Vol. 18, No. 3 (April 1988), pp. 7-10.

41. A particularly thoughtful West German perspective on nuclear weapons in Europe after INF is Volker Ruehe, "Germany Faces Range of Issues After INF," *Wall Street Journal*, February 8, 1988.

42. Robert J. McCartney, "Honecker Proposes Germanys Free of All Nuclear Weapons," *Washington Post*, December 12, 1987, p. A-12; and McCartney, "Soviet Calls for Further Arms Cuts, Shevardnadze Cites Tactical A–Weapons," *ibid.*, January 19, 1988, p. A-10. Chancellor Kohl's rejection of the Honecker initiative is reported in David Marsh, "Bonn Rejects Honecker Ploy on Nuclear Arms," *Financial Times*, April 21, 1988, p. 2.

43. For one prescription on how to modernize while cutting U.S. nuclear weapons in Europe unilaterally by half to 2,300, see Robert D. Blackwill, "NATO: Reduce and Modernize at the Same Time," *International Herald Tribune*, March 1, 1988, p. 4.

44. The most detailed account of that Soviet offensive against NATO INF deployment and of disputes within the U.S. government on INF arms control is found in Strobe Talbott, *Deadly Gambits* (New York: Alfred A. Knopf, 1984).

16

European Security After INF: From Nuclear to Conventional Arms Control

Karsten D. Voigt

For the first time, Europe is facing success in the realm of nuclear arms control. The zero-based solutions of INF have led critics to fear, and overly optimistic elements of the peace movement to hope, for a denuclearized Europe — a widened zero solution. A third zero solution for land-based, short-range missiles, which I would support, would have an impact on the nuclear strategies of the two alliances, but would not necessarily lead to denuclearization. Even if the United States and the Soviet Union withdrew all of their land-based nuclear weapons from Europe, the member states of the two alliances would remain part of the nuclear deterrence strategy of the two nuclear world powers. Although Europeans possess a great deal of influence in reformulating the nuclear deterrence strategy, the final surmounting of nuclear deterrence lies in building a foundation for the step-by-step demilitarization of the East–West conflict. It does not necessarily involve denuclearization or "conventionalization" of Europe, although conventional stability makes sense in this case. Successful steps towards restructuring conventional forces so that they are unable to attack, and towards a demilitarized East–West conflict, are possible only when military intentions are purely defensive in nature. Conventional stability must be founded upon this basis. Détente can remove crises which lead to war and potential animosities in Europe.

Arms control has a chance when its dimensions are seen as disproportionately positive compared to the realities of political intentions based on aggression. In the event that the Gorbachev leadership in the Soviet Union is convinced that the declared objective of the use of Soviet conventional superiority in Europe for political ends is detrimental to substantial advances in the realm of conventional arms control, Mr. Gorbachev should bring it to an end. Should the Soviet Union be prepared to come to an agreement on conventional stability, it would be proof of a new substantial Soviet foreign policy with respect to Western Europe.

Even more important than the details of military planning is a change in the East–West political relationship so that the military element is no longer the dominating factor; its influence must be limited. The demilitarization of relations in Europe and the *repolitization* of the East–West conflict are, for this reason, two sides of the same coin. Arms control is a civilized political peace process where both sides gradually adopt a violence-free method of conflict management. Such a learning process is also required in the discussion of "structural inability" to attack. The debate in itself has changed perceptions. It civilizes, in that it concerns itself with the following problems:

- development of common criterion for military planning;

- communication about and changes in the images of threat;

- orientation of military needs using the principle of *sufficiency* instead of worst-case analyses; and

- conclusive changes in what is meant by standards of *sufficiency* when both sides orient themselves on this principle.

The Asymmetrical Conventional Arms Race

The arms dynamic in conventional weapons in Europe is characterized by a number of asymmetries which are the chief source of instability. They are, at the same time, the major reason why conventional arms control in Europe is faced with obstacles which can be surmounted only with great difficulty. The unsuccessful Mutually Balanced Force Reductions (MBFR) attempt is a primary example. First, there is the geographical asymmetry of the alliance systems. The Warsaw Pact has a cohesive alliance region with enough territorial depth to deploy its troops in deep formation. To move units from western regions of the Soviet Union to the East–West border requires a land journey of less than 1,000 kilometers. On the other hand, Soviet access routes to waterways are extremely unfavorable, most of them usable only with great difficulty or not at all during the winter months. In contrast, NATO has an advantage here and can control access to Eastern European waterways with relative ease. NATO's major handicap, however, lies in the lack of alliance territorial depth. Furthermore, the Atlantic Ocean separates the Western European Allies from a major NATO power, namely the United States. American and Canadian reinforcement units must be transported more than 6,000 kilometers by air and sea in order to reach Europe.

In addition to this geographical asymmetry exist differentiated military doctrines and strategies. In the event of a crisis, the Warsaw Pact hopes to defend itself on its opponent's territory. Its military strategy is, therefore, geared toward a territorial counteroffensive capacity (i.e., forward defense). The development of Soviet strategy is aimed at a comprehensive, conventionalized counteroffensive capacity. NATO fears an offensive following a brief period of preparation much more so than the possibility of the famous *Blitzkrieg* option. Even if no intentions to conduct an aggressive strike can be attributed to the Soviet Union, from a Western point of view, its military strategy still constitutes the greatest source of instability in Europe.

By comparison, the NATO flexible response strategy is admittedly more defensively structured, inasmuch as NATO is in no position to mount a territorial offensive with land units. That having been said, NATO does not renounce the options of bomber and rocket tactical offensives deep in enemy territory. Interventions against military targets in Eastern Europe have always been one of the many NATO air strike unit tasks. In the future, this tactical-offensive capability is to be reinforced under the planning directive, Follow-On Forces Attack (FOFA) and, above all, to be conventionalized. The object is to counter the second Soviet detachment as it advances. In addition, NATO reserves the nuclear first-strike option in its doctrine. This so-called premeditated escalation is being implemented into appropriate weapons systems and operational planning.

Sociopolitical asymmetry also influences the military situation in Europe. While the world power status of the Soviet Union has, hitherto, been based on military might, the Gorbachev reform program suggests medium- and long-term changes in this. The West has far greater economic potential and a much more attractive social system. In connection with this, it should be kept in mind that Soviet forces serve the function of guaranteeing Soviet influence in Eastern European security and social policy. Military forces compensate for a lack of acceptance enjoyed by the political system. Systemic differences are also reflected by differing degrees of openness and availability of military planning to the public on both sides. In the conventional sphere, these asymmetries undoubtedly influence the dynamic trends in armaments. The Soviet Union and other Warsaw Pact nations are gradually stepping up and conventionalizing their potential, which is suited for a territorial offensive. Conversely, there are also tendencies in the West which could be referred to as a "Sovietization" of its military doctrine. This is to be found in the conventionalization and shift in capacity towards an emphasis on tactical offensives (e.g., FOFA and AirLand Battle). On both sides, trends in arms technology match these plans for the further consolidation of offensive options.

The long-range potential of conventional munitions is being increasingly extended. Both sides are striving to acquire ever greater capacities for the rapid deployment of military resources to enemy territory (e.g., conventional stand-off weapons, cruise missiles, and, to some extent, ballistic missiles). Certainly, the technical and financial problems which arise as a result of these new technologies are not to be underestimated. Because of tendencies both sides exhibit in arms technology and military strategy, it should, nonetheless, further magnify existing instabilities, unless limitations can be determined and agreed upon.

Criteria for the Stabilization of Conventional Armaments

An increasingly violence-free management of the power and systemic conflict between East and West could also have a humanizing effect, as defined by the CSCE Final Act, on the systems and, thus, on the substance of conflict. There is little evidence to suggest that the West has an overall concept of peace and security policy, a failure which becomes increasingly obvious with each new Soviet disarmament proposal.

In the suggested conventional stabilization framework, both sides need to place more emphasis on the creation of a nonaggressive force structure. For the Warsaw Pact, this would mean permitting its military territory-taking counteroffensive strategy to be called into question and, ultimately, abandoning this provocative concept. NATO's response would be to reverse its trend towards the conventional consolidation of airborne capability to strike Soviet forces. Put simply, if the Warsaw Pact were to abandon its strategy which provides for a territory-taking counteroffensive, NATO's threat to exercise a nuclear first use would lose any military justification.

The structural inability to attack is unquestionably a more demanding goal than that of mere parity. If the military power ratio is compensated by arms control policy, under the traditional ideas of balance, a contribution to the goal of conventional stabilization will already have been made. On the other hand, a defensive force structure requires both sides not only to disarm but, above all, to rearm within its force structure such that it expresses a credible and clearly recognizable renunciation of offensive options. In this sense, both sides would qualify for parity in the direction of defensive capabilities.

Conventional stabilization focusing on the structural inability to attack should be based on the following criteria:

- Strategic stability — In the conventional sphere, this means that both sides have force structures which facilitate defense but make attack more difficult. In this context, confidence and security-building measures (CSBMs) serve the purpose of reducing surprise attack capability, as well as the ability to conduct an attack after a short period of military preparation. In addition, for both sides to reduce their offensive options, these potentials must be limited and/or converted, and military doctrines, as well as strategies included in the East–West dialogue, must eventually be changed.

- Stabilization of arms dynamics — This concerns restrictions which affect military power relationships. The goal should be for both sides to create defensive structures which do not further generate stimulation of the arms race.

- Crisis stability — If force structures produce no pressure for the rapid deployment of military resources, it makes the race to mobilize unnecessary, places a premium on winning time, and is considered crisis-proof.

CSBMs, which lengthen periods of military preparation, such as disengagement measures, can be cited in this context. In the final analysis, however, only direct limitation and alteration of potentials can create true crisis-proof structures.

- Restraint capacity — Even after adjustments towards structural inability to attack are made, any military aggression must continue to be calculatedly untenable. The capacity to exercise restraint must be geared to denying the chances of the success of an attack (criterion: denial concept of deterrence), in terms of conventional offensive potentials. On the other hand, the military role of nuclear weapons is to change; they will only be a last-resort option.

- Limitation of damage — Because of the primacy of the prevention of war, the four criteria above need to take precedence over that of damage limitation. But, as failure to prevent war cannot be completely ruled out, the structures of armed forces must provide no compulsion towards rapid escalation in the event of war. Damage limitation also entails the protection of that which is to be defended — no easy task in densely populated Central Europe. Even a long drawn-out conventional war would lead to the destruction of Central Europe.

- Alliance acceptability — For NATO, this means retention of its multinational forward unit structure. This applies to both restraint capacity (an aggressor would have to contend with the entire Alliance) and stable policy considerations (no NATO nation can single-handedly attempt an offensive). Conversely, for the Warsaw Pact, alliance acceptability means the creation of force structures which facilitate a more organic and less hegemonic relationship between the Soviet Union and the Eastern European nations.

- Social acceptability — The defense instrument must be capable of being brought into conformity with the standards of democracy. That is, it may neither lead to the militarization of society, nor may the financial and human resources of society be excessively strained so that other state tasks (such as social, environmental, and educational policy concerns) cannot continue to be carried out.

Options for Conventional Arms Control in Europe

The history of conventional arms control in Europe presents, thus far, a contradictory picture. The first attempt at conventional stabilization has been unable to produce a result; *even if the MBFR negotiations had been successful, the military situation in Europe would scarcely be altered.* The positive and negative experiences acquired during MBFR negotiations can be expanded upon and utilized in preparation for a broader European disarmament conference. It is my view that the MBFR negotiations should not be abandoned until the chances for substantive agreements at a European conference are seen to be greater than those of the

MBFR negotiations. An extension of the MBFR zone to include Denmark and Hungary would, from an arms control perspective, provide a sensible Central European subregion within a European disarmament conference. The balance sheet to date in the area of confidence-building measures is much more positive. The Stockholm agreement of the Conference on Confidence and Security-Building Measures and Conventional Disarmament in Europe (CDE) constitutes, from a perspective of security policy, the first step towards improving military attack preparations warning signals. Although the agreement contains gaps — for example, it permits emergency alert exercises without prior warning — it does somewhat reduce the risk of surprise attacks. The creation of permanent pan-European institutions, which could assess the already agreed-upon confidence-building measures and broaden further ones, would foster this process.

Priority must now be given to the new conventional arms control negotiations. It was General Secretary Gorbachev who proposed fresh talks concerning conventional reductions from the Atlantic to the Urals in East Berlin on May 18, 1986. In doing this, he theoretically accepted the old Western claim that stabilization of the European military situation requires that western military regions of the Soviet Union be included.

The Warsaw Pact Budapest Appeal of June 11, 1986, elaborated upon and further defined the Soviet proposal. It called for negotiations of all components of land and tactical air forces, as well as battlefield nuclear weapons from the Atlantic to the Urals. The reduced forces would be disengaged as complete and equal major units, both intermediate troop formations and units and their weapons. To obviate the danger of surprise attacks, the concentration of troops along alliance borders would be reduced. The new conventional arms control discussions were to be held within the CDE. In relation to verification, the Warsaw Pact stated that it would accept on-site inspections. Lastly, the Budapest Appeal contained one very attractive indication regarding the conventional stabilization discussions; both sides should base their military doctrines on "defensive principles."

Apart from this vague hint in the Budapest Appeal, there are other indications that the Warsaw Pact is prepared to question its own military doctrine:

- The SPD–SED agreement of October 21, 1986, concerning a nuclear-free corridor, calls for a reduction in the nuclear and conventional strike capacity of both alliance forces. The corridor agreement provides for the mutual withdrawal of all artillery carrier systems capable of carrying nuclear weapons and, thus, of all heavy conventional artillery.

- In his Prague speech of March 12, 1987, Mikhail Gorbachev announced his readiness to remove the elements of imbalance, and asymmetry, should they exist. Reductions would have to be initiated by the party who enjoys the advantage. As previously, the Soviet Union now maintains that, all in all, there is a roughly balanced power relationship in Europe. By adapting the Western concept of asymmetry, however, Gorbachev intimated a cautious turn which Soviet authors have long avoided.

- Attention should be drawn to the latest Polish disarmament plans. In addition to the familiar elements, the Jaruzelski initiatives stress that NATO and the Warsaw Pact should alter their military doctrines such that their respective strategies can be recognized as strictly defensive. Also significant in the Polish plans is the zone in Central Europe — which, in contrast to MBFR, includes Denmark and Hungary as well as the Federal Republic of Germany, the Benelux countries, the German Democratic Republic, Czechoslovakia, and Poland — designated for nuclear and conventional reductions. Such reduction zones were first dealt with in great detail at the Socialist Party of the FRG (SPD), the Socialist Unity Party of the GDR (SED), and the Communist Party of Czechoslovakia (CPC) discussions. They were set forth by the aforementioned parties in a common declaration from Prague on May 14, 1986.

- The East Berlin Summit meeting demonstrated Warsaw Pact preparedness to consult both sides concerning military doctrine and repeated the Budapest Appeal's call for the adoption of defensive principles. In lieu of existing historical and geographical asymmetries of both sides in Europe, the Eastern alliance simultaneously exhibited its support of renouncing those elements which constitute military imbalances. The respective party would be required to abolish that which provides strategic advantages.

Thus, the Warsaw Pact nations have, in principle, accepted structural inability to attack as the goal of European conventional arms control. Moreover, the Soviet Union has indicated that it is prepared to take European security interests regarding medium-range nuclear weapons more seriously than in the past. Exactly how seriously remains to be seen in the ultimate conventional stabilization. Over and above statements of principle, the signs of Eastern reevaluations in the realm of conventional weapons must also lead to concrete actions.

Above all, it is important that the West now find its way to a consistent arms control strategy. It should be noted, however, that NATO reactions to recent Warsaw Pact proposals indicate that the alliance suffers difficulty in trying to formulate a common position. This may, in part, reflect feelings of a conventionally inferior questioner but is, nonetheless, inexcusable.

The compromise reached at the NATO meeting in Reykjavik brought about a common Western position on the definition of the negotiating framework. The Western goals in these negotiations still remain unclear.

So far, NATO has been unable to conclude a negotiating concept. Over the years, there apparently has been insufficient consideration in Brussels, as well as within the individual NATO states, of conventional arms control. Within the alliance there is no apparent concept of demands that should be placed upon the Warsaw Pact, nor what structural changes the Soviet military strategy must initiate, let alone what concessions, if any, the West may proffer the Warsaw Pact. In doing this, the West is open to suspicions that it is not sincerely interested in the stabilization of asymmetrical structures for itself in the sphere of confidence building. As important as they may be, confidence-building measures are insufficient if conventional stability is seriously desired; cuts and changes in military potential cannot be avoided.

Options for Conventional Arms Limitations

Arms control negotiations must, as mentioned earlier, be aimed at the goal of mutual structural inability to attack in the sense of the aforementioned criteria. The MBFR endeavor is inadequate to serve this purpose because it focuses on troop levels and because the room for reductions is, for well-known reasons, too narrow. Conventional arms control must be aimed at mutual reductions in attack potential after a brief preparation and at the elimination of the peak periods of offensive threat.

In an initial phase, European conventional arms control negotiations should concentrate on land-based weapons and forces. As a matter of negotiating technique, the Navy should be excluded from the negotiations. The Air Force should only be included in the first phase insofar as a settlement between renunciation of Warsaw Pact offensive land-based options and NATO offensive air-based options (e.g., FOFA) seems sensible and necessary.

Although the room for reductions at MBFR is too narrow, it continues to appear sensible to designate at least one subregion within the pan-European negotiating zone from the Atlantic to the Urals (the Central European zone) in which the main concern is reductions and restrictions, and the rest of the territory from the Atlantic to the Urals as one in which all troop movements and activities would be restricted. An agreement regarding reductions in the Central European zone should be formulated by all NATO and Warsaw Pact nations it would effect. Participation of all CSCE nations in the verification of rules could link this agreement with another concerning the pan-European negotiating zone. As the Polish proposal envisages, the central reduction area should cover Denmark and Hungary in addition to the MBFR zone.

It becomes increasingly urgent that France outline its intentions in the realm of conventional arms in Europe so that it is possible for France to actively participate in negotiations. It was the French government who, in the 1970s, originally proposed "from the Atlantic-to-the-Urals" arms control talks. The more concrete discussions become, the less clear it becomes where France, as a Western European union and nonmilitarily integrated NATO member, is prepared to commit itself to the disengagement of forces and arms.

Despite detailed assessments of the power ratio between NATO and the Warsaw Pact (the current debate, which juggles figures such as 1:3, 1:2, or 1:1.2, misses the real problems of such power comparisons by a large margin), the Warsaw Pact will have to submit to asymmetrical reductions which are at its disadvantage. The primary source of military instability in Europe is, after all, the Soviet territory-taking counteroffensive strategy. Yet, it is prepared to accept asymmetrical reductions if the West is willing to alter its own strategy (for example, NATO renunciation of the introduction of new conventional FOFA options in the future). Here again, it is a matter of taking Gorbachev at his word.

In the final analysis, offensive options can be dismantled only if capabilities are genuinely reduced. Military depletion and movement restrictions, however important they may be, must not come to the fore when mutual structural stability is at stake. In order to limit strike options after a brief preparatory period, the reductions should be concentrated on combat-ready active units, as regards ground forces, with more than 50 percent present (this would include Soviet category I and II units). From the outset, these unit reductions must include weaponry. In addition, material reductions should be concentrated on mobile weapons with high firepower capabilities which serve strike options (tanks, heavy artillery, combat helicopters, and, in particular, tank guns).

To combat asymmetries, combat-ready unit reductions and that of their weaponry could be proportioned in accordance to contributions of individual members of the respective alliances. The Soviet Union would then have to bear the lion's share of Warsaw Pact reductions, whereas NATO reductions would be more uniformly distributed. Senator Sam Nunn (D-GA) has proposed that the United States and the Soviet Union withdraw half of their troops stationed in Western and Eastern Europe. For the U.S., that would be two-and-a-half divisions, but for the Soviet Union, thirteen and a half. If these divisions are not dissolved, but merely withdrawn from Europe, problems arise in respect to geographical asymmetry. In the event of a serious situation, the Soviet Union could transport its troops to Eastern Europe more rapidly than the United States could send its units to Western Europe. One way to overcome this complication might be to allow the Americans to store pre-positioned materials (POMCUS stores), provided that such storage did not conflict with the aforementioned agreed limitations on offensive equipment.

In addition, there should be accompanying limitations which first serve as confidence-building measures and, concurrently, reduce the risk of attack after a brief military preparation period:

- mutual transformation of combat-ready troop units into reduced strength units (agreed strength reduction);

- selective reduction of armored units;

- depletion zones in which the storage of materials suitable for attack, such as tanks, combat helicopters, heavy artillery, munition and fuel depots, and bridge-building equipment, is forbidden; on the other hand, defensive equipment such as barrier mines and tank barriers would be permitted in this zone; and

- more rigorous CSCE provisions, particularly that an obligatory advanced warning should be given for emergency exercises of a certain size.

These proposals ought to indicate the direction in which discussions should proceed in order to advance conventional stability in Europe. The mutual creation of instruments which correspond to the structural inability to attack is an ambitious long-term goal which can only be achieved at a gradual pace. Thus, it is all the more necessary for both sides to initiate the requisite learning process of comprehensive security and the disarmament dialogue between East and West allowing nonaligned and neutral states to participate. For this dialogue to achieve substantial results in formal governmental negotiations, it must be scientifically prepared and substantiated. Politicians, parties, and parliaments can encourage the development of a pan-European security and disarmament policy in their discussions, which, given superpower participation, would assist in dismantling hostile images and enmity in the East–West conflict. Backed by proposals, they can improve the chances of success of these intergovernmental negotiations.

The advancement of arms control negotiations does not emerge as the result of political tactics, nor from domestic or alliance political pressures. It is stimulated by political problems which arise in the application of alternative defense concepts. Without the enemy's cooperation, without the introduction of changes in the military structure and a mutual concept of détente, conventional stability will never be achieved. This can be verified by the Warsaw Pact's inability to determine what constitutes a nonthreatening and confidence-building security structure in Europe.

The disagreements between arms control supporters and followers of unilateral rearmament can be overcome when both images are tied together by one proposal for conventional stability in Europe. The exclusively defensive rearmament should not become dependent on the results of arms control negotiations. Nor should one submit to the illusion that a long-lasting, stable structure could be created in Europe in the absence of a cooperative initiative to reorganize the present status of armaments.

Through a rearmament, arms control negotiations and unilateral stabilization could both complement and challenge one another. It would be the responsibility of negotiators to effect Warsaw Pact reductions. Negotiations concerning the minimization of conventional armament in Europe should begin with the goal of mutually dismantling mobile and heavy armed forces. From there, the talks can determine acceptable criteria of stability and discuss threat assessments. In any case, they must mutually set forth a definition of sufficient defense capacity. It then becomes a two-sided learning process of how to coordinate conventional stability.

Defense politicians have the responsibility of managing the establishment of the defense structure. As long as sufficient defense capability is not placed at risk, stabilization can begin to develop independently of arms control results. The limits must be defined, lest its restraint capacity no longer be guaranteed. Based on its enormous potential stationed along the borders, the Soviet Union has, by far, more room to take unilateral steps in Europe than does NATO. A truly decisive step would be for the Warsaw Pact to renounce its forward strategy and establish one of forward defense. This presumes changes in both the doctrine and structure of the Warsaw Pact forces, which would provide NATO with room to take unilateral steps.

The potential for such an integrated, reform-oriented strategy to succeed in advancing the military situation in Europe has improved. The indications that Gorbachev's new thinking has stimulated a new thought process throughout the Warsaw Pact are increasing. The chances that this should further develop must be tested by corresponding initiatives from the West.

17

The Soviet Union and Conventional Arms Control: What Lies Ahead?*

David B. Rivkin, Jr.

For over two years, the prospects for the new "Atlantic-to-the-Urals" arms control talks between NATO and the Warsaw Pact have been a much discussed subject among Western analysts and newspaper commentators. While expert opinions on this subject differ widely, ranging from euphoria to an outright skepticism about Soviet intentions, there appears to have emerged a general feeling that a major breakthrough in conventional arms control, capable of transforming the post-World War II posture of military stand-off between the two opposing alliances, is possible. To a large extent, this feeling has been fostered by a perception that, under the leadership of Soviet General Secretary Mikhail Gorbachev, Soviet diplomacy has been imbued with a new dynamism, and dramatic reverses in erstwhile Soviet stances on virtually any issue cannot be ruled out.

To be sure, the history of conventional arms control to date provides no reasons for optimism. The Mutual and Balanced Force Reduction (MBFR) talks, now in their fifteenth year, have removed not a single American or Soviet soldier, sailor, or airman from Central Europe. Since the talks began in 1973, the Soviets have actually increased the size and firepower of their ground forces forward deployed in Eastern Europe. During this entire period, it has never appeared to be in Moscow's interest to give up or trade away its conventional force superiority in Europe, a superiority reinforced by profound geographic advantages (i.e., the U.S.S.R.'s proximity to Central Europe and NATO Center's lack of depth). And it would seem that Moscow would have an even stronger interest in preserving that

---

This paper is based in part on a recent Foreign Affairs *article entitled "Defending Post-INF Europe," which he co-authored with Jeffrey Record.*

superiority in a post-INF Treaty environment, characterized by prospective unilateral national force reductions within NATO attributes to a host of budgetary, political, and demographic pressures which face the alliance members.

As of late, however, some Western analysts have argued that Gorbachev, having already taken bold steps in the nuclear arms control area, is bound to move eventually toward stabilizing reductions in conventional forces. The apparent assumptions behind this view are that Gorbachev wants to reduce Soviet defense spending, that he is not really concerned about the threat of war, and that, hence, he does not take Soviet warfighting requirements seriously. It has been also argued by such analysts of the Soviet military as Michael MccGwire that, since the Soviets no longer believe that an East–West conflict in the Third World would inevitably escalate or spill over into Europe, they are beginning to accept the desirability of a defensive posture in Eastern Europe.

Yet, there is little hard evidence to support these judgments. Unless one assumes that Gorbachev is seized with an across-the-board arms control fever, Soviet attitudes toward stabilization of the conventional force balance in Europe cannot be simply extrapolated from recent Soviet theater nuclear arms control decisions. More specifically, it is most unlikely that Moscow, having decided to forego some of its theater nuclear warfighting options, is now prepared to relinquish its conventional warfighting options as well. No matter how low the probability of war in Europe may be in Soviet eyes, it is hard to believe that Moscow would now be ready to rely indefinitely on a deterrence-only posture, with no provisions made for failed deterrence. It is also difficult to see what inducements there are for Moscow to give up its offensive conventional warfighting options, since NATO has no offensive options of its own which it can exchange in trade.

Soviet Force Structure and Doctrine

The history of Soviet conventional force sizing decisions fully supports this conclusion. Even Khrushchev, who was prepared to consider dramatic cuts in Soviet conventional forces and who did implement some significant force reductions, was ready to do so largely because of his conviction that the ongoing revolution in military affairs made traditional ground warfare obsolete, and that Moscow would eventually acquire a potent nuclear warfighting posture.

Arguably, Khrushchev was also prepared to postpone procurement of some strategic weapons until more capable systems became available, on the theory that the imperialist threat could, for the time being, be held at bay by nuclear saber rattling and shrill diplomacy. Even this approach of temporarily foregoing the optimization of Soviet defense capabilities was severely criticized by Khrushchev's successors. There is no evidence to suggest, however, that either Khrushchev or his successors were ever willing to go so far as permanently to give up the search for credible warfighting options as a means of bolstering deterrence, safeguarding Soviet national security should deterrence fail, and providing potent political leverage for peacetime Soviet foreign policy. Some Western commentators nevertheless believe that a more restrained Soviet foreign policy prompted by the "new thinking" in Moscow will sooner or later result in a scaling down of Soviet military power and aspirations. This judgment merits close scrutiny, because there is evidence of emerging changes in the Soviet military planning and force posture, especially in the area of nuclear weapons policy.

The emphasis accorded to nuclear weapons on Soviet military planning and the specific roles assigned to them have, of course, changed over the last three decades. In the early 1960s, Soviet doctrine postulated, in the event deterrence failed, prompt escalation of any direct East–West conflict into an all-out war in which nuclear weapons would be used massively and with decisive results. By the late 1960s, the Soviets began to acknowledge the possibility of a conventional phase in a world war, and claimed that escalation, while still likely, was not inevitable. In the early 1970s, Moscow seems to have flirted briefly with the idea of a separate theater nuclear option, to be exercised under the aegis of strategic stalemate.

By the late 1970s, however, Moscow rearticulated the concept of the inevitability of prompt escalation stemming from any Western use of nuclear weapons against targets on Soviet territory, and began to place greater emphasis on conventional operations. Indeed, the Soviets began to discuss the possibility of keeping even a major East–West conflict entirely conventional, a development reflected in Soviet adoption of an unconditional nuclear no-first-use policy and the staging of several all-conventional military exercises. In 1977, Brezhnev publicly renounced the value of nuclear superiority, and by the early 1980s, the feasibility and the desirability of seeking victory in a nuclear war had lost official sanction.

Apparent Soviet disillusionment with nuclear warfighting has not, however, been attended by any attraction to the concept of Mutual Assured Destruction (MAD). Soviet writers claim with increasing frequency that the "balance of terror" creates mutual suspicion, provokes tensions and conflicts in international relations, and stimulates the arms race. MAD may be temporarily acceptable, but offers no long-term solutions to Soviet security needs.

Though the issue of whether Soviet rejection of traditional nuclear emphasis represents a genuine change in Soviet planning and force procurement — or is merely a ruse — has provoked a talmudic debate among Western defense experts, it is enough to say that even a sincere and complete Soviet rejection of nuclear warfighting does little if anything to alter the threat posed by Moscow's conventional military superiority on the Eurasian land mass. Moreover, Soviet force procurement trends suggest a determination to retain the necessary flexibility to implement the full range of nuclear options.

The idea that the apparent Soviet disillusionment with the political and military utility of nuclear weapons has not resulted in discernible changes at the operational level of Soviet nuclear weapons policy may be a bit puzzling to Western defense experts. After all, in the West, the growing popular enchantment with the fashionable precepts of denuclearization has greatly inhibited the modernization and deployment of new nuclear weapons. The Soviet military, however, remains convinced that, while a nuclear war would be an unmitigated disaster and ought to be avoided at almost any cost, they ought to retain nuclear warfighting options in the years ahead. The reasons for this attitude are quite simple — Moscow believes that, while NATO or U.S. resort to nuclear weapons in an East–West conflict is most unlikely, it cannot be ruled out. The Soviets are also concerned about the potential spread of nuclear weapons and ballistic missiles to third countries and growing nuclear arsenals of such countries as China; and last, but not least, in their view, nuclear weapons offer the means to terminate, on favorable terms, a conventional war gone awry. Overall, the Soviets believe that a robust nuclear force posture constitutes a good insurance policy, and they are not about to throw it away.

Not surprisingly, the Soviets continue to modernize their nuclear forces, improving the accuracy of their ballistic missiles and lowering the yields of their nuclear warheads to reduce collateral damage. They are also developing replacements for the SS–24 and SS–25 ICBMs that are only now beginning deployment, as well as replacements for their new SS–N–20 and SS–N–23 SLBMs. New SSBNs are also being built. By the mid-1990s, the Soviet strategic inventory will consist almost entirely of these new weapons, of which some forty percent will be on mobile launchers. The Soviets are further reported to have tested chemical-biological warheads on their ICBMs, including the SS–19. Meanwhile, the Soviet bomber force is undergoing rapid modernization via deployment of new highly capable aircraft, endowing Moscow with flexible intercontinental and theater air attack capability. These activities, combined with on-going Soviet development and production of the elements of a country-wide ballistic missile defense system, the ABM–X–3, clearly indicate Soviet willingness to invest the necessary resources in retaining some options, however imperfect, of prevailing in a nuclear conflict.

In the conventional force arena, the Soviet military is engaged in a major restructuring of ground and tactical air forces, prompted in part by a relative deemphasis on divisional-level warfare. The Soviets evidently plan to confer greater autonomy and initiative upon small unit commanders, and to develop smaller, leaner, and better trained forces. For the last several years, Eastern European militaries have been serving as laboratories for testing various Soviet operational innovations. This shift in approach to conventional warfare may also enable Moscow to implement some cutbacks in defense spending and is fully compatible with the concept of reasonable sufficiency embraced by the top Soviet political and military leaders. The Soviets are also looking toward obtaining a capability for an immediate massive attack by aircraft and conventionally armed ballistic missiles in the event of war — an air operation designed to obliterate key NATO targets heretofore allocated to nuclear weapons.

Supplementing this capability are major innovations in Soviet operational methods for dealing with NATO's defenses. In place of traditional Soviet operational rigidity and reliance on complex, prescribed attack plans for breaking through NATO's lines, Moscow is beginning to emphasize flexibility and opportunistic attack scenarios. Thus, newly constituted Soviet operational maneuver groups and other mobile formations are designed to flow through NATO forward defenses rapidly and disrupt the alliance's "soft" rear area. Last, but not least, the

Soviets are planning to use *Spetsnaz* and other unconventional warfare units to destroy NATO nuclear weapons sites, command and control centers, and "smart weapons" early in the conflict.

And even more radical doctrinal innovations are underway. The Soviet military appears to be seriously considering the advisability of a major restructuring of their military forces arraigned against NATO. The former Soviet approach has been to cram as much military capability forward as possible, enabling the Warsaw Pact, in case of war, to break rapidly through NATO defenses and seize NATO territory with such speed and decisiveness, as to, hopefully, prevent NATO from resorting to nuclear weapons. The Soviets, however, began to realize that potential proliferation of long-range strike systems in the NATO arsenal makes forward deployed Soviet forces vulnerable to destruction. Moreover, the manifestly offensive emphasis of Soviet forces has been of such concern to NATO that it has provided a major impetus for Western military efforts. In short, the Soviet force posture in Europe has become questionable from both political and military standpoints.

To cope with this problem, the Soviet military has begun to consider the possibility of conducting a different kind of war. If Moscow thinned out its forward deployed forces, it could concentrate during the first phase of a conflict on destroying NATO airpower, long-range strike weapons, and nuclear systems. During this phase, Soviet ground forces would remain dispersed and eschew large scale offensive operations. Once NATO's airpower and strike assets have been attrited, the Soviet ground forces would go on the offensive. At that time, they also would be able to mass more readily and, thus, bring a greater combat weight to bear on NATO defenses.

In conjunction with rethinking overall force posture, the Soviets have been also exploring the possibility of evolving a force mix featuring a more balanced composition of armor, mobile infantry, and strike systems. Another key area of exploration has been the emphasis on developing forces capable of greater endurance and versatility and trained to execute both offensive and defensive operations. In many instances, the Soviets have already progressed beyond theoretical discussions, introducing actual changes into their troop training, exercises, and even force structure. In the latter category, for example, there has been evidence that Moscow has been experimenting with the introduction of such new types of formations as air assault brigades and unified army corps.

All of this amounts to further growth, not diminution, in the Soviet conventional threat to NATO. In sum, there is nothing in the way of visible changes in Soviet military doctrine and force structure that offers any comfort to NATO.

Arms Control Prospects After INF

What, then, of prospects for a negotiated reduction in the Soviet threat to NATO? Is there reason to hope that the INF Treaty will be followed by meaningful conventional arms control in Europe?

Unquestionably, Moscow remains interested in European arms control, and Western European considerations have increasingly influenced the Soviet arms control agenda. Thus, Moscow has sought to tailor its arms control proposals to make them especially palatable to European audiences and to foster an image of the Soviet Union as peace-loving and progressive, sensitive to European concerns, and willing to contribute to the relaxation of tensions in Europe. Moscow is clearly interested in improving its relations with Western Europe at the expense of the United States and has long viewed arms control as a useful tool in accomplishing that objective. The prevailing Soviet view seems to be that arms control inherently strengthens international security and benefits Moscow.

Yet, Moscow also appears to recognize that earlier judgments of Western Europeans as overeager for arms control deserve qualification. The Soviets were apparently jolted by negative European reactions to both President Reagan's "denuclearization" proposals presented in Reykjavik and the "double-zero" INF deal. Reportedly, the Soviets also were impressed by the vigor with which Margaret Thatcher defended the deterrent value of nuclear weapons during her trip to Moscow. At the same time, however, Moscow undoubtedly noted that the Europeans, their misgivings aside, were unable to reject the double-zero deal. Chancellor Helmut Kohl's reluctant agreement to scrap Germany's 72 Pershing 1A missiles also has not gone unnoticed in the Kremlin.

The Soviets, in short, seem to be taking a more realistic view of how much can be accomplished in arms control by appealing to Western public opinion. The failure of heavy-handed Soviet diplomacy to derail NATO's original INF deployment through pressure on Western European governments and peace movements revealed the limits of what "revolutionary forces" can do to influence Western European policy. Moscow also appears to realize that the U.S. Congress, while it can be a useful source of pressure on the Reagan Administration, cannot be counted upon to favor policies the Soviets seek to promote. Thus, while Gorbachev appealed to Congress in the aftermath at Reykjavik, Soviet commentators have subsequently noted congressional opposition to the idea of abolishing all nuclear weapons. Likewise, Moscow concluded that its efforts in 1983 to whip up a war scare in Europe by engaging in such dramatic gestures as the walk out from the Geneva arms control talks failed to produce the desired effects.

The question still remains as to whether Moscow is or will be prepared to follow up the INF Treaty with serious efforts to obtain a comprehensive conventional force agreement. The answer to this question hinges on Gorbachev's own national security agenda. A not-unlikely scenario is that following the INF Treaty's ratification by the U.S. Senate, Gorbachev will launch a conventional arms control offensive. Undoubtedly, any new Soviet proposals would be presented in a fashion designed to appeal to Western public opinion. Moscow can also be expected to lean on individual Western European countries to secure their agreement to Soviet arms control proposals and to threaten to hold their respective bilateral relations hostage to the good behavior of the countries involved. The Soviets are further likely to seek to capitalize on West Germany's growing, if vague, yearning for reunification and on Germany's anxiety about its unique vulnerability in a post-INF NATO. For example, Moscow, despite persistent German requests over the last several years, has refused to schedule a Gorbachev–Kohl meeting.

Some of the specifics of Moscow's anticipated conventional arms control package are likely to reflect the substance of recent Soviet proposals. In April and June of 1986, the Soviets proposed implementing deep reductions in military forces in Europe and replacing or supplanting the moribund MBFR discussions with new talks on conventional arms that would cover the entire geographical area stretching from the Atlantic to the Urals. In 1987, NATO agreed to the proposed concept of new negotiations and began the so-called mandate talks to determine the parameters of these negotiations. Concurrent with the new talks, the confidence-and-security-building measures and disarmament in Europe (CDE) process is likely to commence as well, picking up where the 1986 Stockholm accords left off.

The specific Soviet proposals contained in the so-called Budapest appeal issued during the June 1986 Warsaw Pact meeting were as follows: reductions would apply to ground forces and tactical air forces of European participants, as well as to the corresponding forces of the United States and Canada stationed in Europe; initial reductions would consist of 100,000–150,000 men for each alliance; second-phase reductions would cut remaining alliance force deployments by another 25 percent. Third-phase reductions would deal not only with the members of NATO and the Warsaw Pact, but also with the European neutral states.

More recently, the Soviets floated a somewhat modified version of their earlier arms control proposal, wherein during the first phase, both sides would concentrate on establishing a mutually acceptable and verifiable data base on the presently existing NATO and Warsaw Pact forces. Once this objective is accomplished, reductions of up to 500,000–550,000 men would follow in the second phase, with still deeper cuts to come later.

These Soviet proposals would seem to preserve the present military balance in Europe, which favors the Warsaw Pact, although at lower force levels. There is at present no reason to believe that Moscow would be prepared to accept an agreement depriving the Pact of its longstanding numerical and geographical advantages over NATO, although Gorbachev, in addition to restating the April 1986 proposal, might choose to play to the galleries of Western European public opinion by some dramatic opening gambit. He might, for example, offer to withdraw unilaterally a few Soviet divisions from Czechoslovakia, or Hungary, or even the GDR, with further reductions conditional on an appropriate NATO response. Removal of, say, three or four divisions could be undertaken without significantly diminishing Soviet offensive capabilities against NATO, especially given the likelihood of a decline in forward-deployed NATO forces and the ongoing restructuring of Soviet ground forces. Curiously, however, despite the persistent speculation about unilateral Soviet withdrawals of some forces from Eastern Europe, in the last few months, the Soviet spokesmen have gone to considerable lengths to deny that this will ever take place.

The Soviets have also indicated that they are prepared, allegedly in deference to longstanding Western wishes, to concentrate on reducing not only conventional forces across the board but also on cutting certain "destabilizing" categories of forces. Thus, a recent edition of the Soviet publication, *Whence Threat to Peace*, argues that reductions in aircraft and armored forces should be given particular priority. Specifically, the Soviets have tentatively proposed a scheme of reductions in European-based aircraft, claiming that aircraft are the most destabilizing forces of all and that, in modern time, aggressive wars were always begun by a surprise air attack. Airpower, however, has been traditionally perceived as NATO's conventional trump card, capable of neutralizing at least some of the Warsaw Pact's superiority in ground forces. Thus, even equitable and verifiable reductions in tactical airpower would, on balance, be to NATO's disadvantage. If such reductions were applied to all NATO countries, there is also a very real risk that smaller national air forces might be particularly hard hit. Any schemes which involve withdrawals of aircraft from Central Europe would also be costly and difficult to implement, for though the Soviets could rebase their aircraft east of the Urals, it is unclear where NATO could rebase its airpower.

Moreover, since, in the wake of the INF Treaty, aircraft will be NATO's sole land-based long-range nuclear delivery systems, the Soviet focus on them undoubtedly reflects a drive to further denuclearize Europe. The Soviets have also made it clear that removing NATO's tactical nuclear weapons and stopping the implementation of NATO's nuclear force modernization plans remains a top priority. While Foreign Minister Shevardnadze, in a recent speech, expressed willingness to postpone temporarily cuts in remaining nuclear forces, he also made it clear that Moscow envisions a clear linkage between conventional reductions and further denuclearization of Europe. The Soviets are also likely to renew past attempts to open discussion on the respective military doctrines of NATO and the Warsaw Pact, with the aims of scoring propaganda points and of curtailing, if possible, NATO's comparatively large-scale military exercises. Such exercises are far more crucial to NATO than corresponding Warsaw Pact exercises are to Moscow.

Another key Soviet priority appears to be the imposition of constraints on Western naval power in general, and American naval forces, in particular. As argued by numerous Soviet military officials, from the Chief of the Soviet General Staff Marshal Akhromeyev on down, no deep cuts in ground forces in Eurasia are possible without the imposition of constraints on Western power projection assets, of which naval forces are the most prominent. Another sentiment, readily discernible in Soviet military writings and statements, is the view that, given the complexity of the issues involved, the conventional arms control process to tackle them would be a long one indeed.

Specific problems with early Soviet proposals and the inherent complexity of arms control discussions, which potentially entail complicated linkages and trade-offs among different types of ground, air, and naval forces aside, there are several fundamental reasons to question the prospects of success in the Atlantic-to-the-Urals talks. To begin with, the Warsaw Pact enjoys considerable conventional superiority in Europe, and Moscow is fully aware of the fact that NATO cannot expand its active-duty conventional forces — indeed, that those forces are likely to shrink. Moreover, to conclude that Gorbachev is facing considerable economic pressures does not necessarily mean that the Soviets are ready to settle for genuine parity in conventional military capabilities. After all, they can unilaterally reduce the size of their forces, cut defense spending, and still retain a range of warfighting options against NATO.

Eastern European political considerations also impose major constraints on Gorbachev's ability to pull Soviet forces out of Europe. Thus, any Soviet reductions in forward-deployed forces would have to be modest in size and gradual. Overall, it is likely that the Soviets, for the foreseeable future, will merely seek to "repackage" their military superiority in Europe. Robert Blackwill, the former U.S. representative to the MBFR talks, has observed that "to believe that Gorbachev will rescue the West from its conventional inferiority is to be on the lookout for Santa."

There is, too, the fact that the anticipated Soviet conventional arms control offensive will fall upon a vulnerable and divided NATO. The alliance has yet to grapple with the INF Treaty's specific long-term implications for conventional deterrence and defense in Europe. Moreover, while there is general agreement within NATO that the treaty's substantial denuclearization of theater deterrence calls for a compensatory reinvigoration of the alliance's conventional defenses, national demographic, political, defense budgetary, and force deployment trends on both sides of the Atlantic portend a worsening of the nonnuclear military balance on the continent. This raises the prospect of negotiations being used as an alibi for not implementing needed conventional force improvements.

Additionally, the arms control process itself, far from easing NATO's predicaments, contains a number of potential pitfalls for the alliance. Attractive Soviet arms control proposals are likely to strain the cohesion of the alliance and to heighten tensions between conservative NATO governments and opposition parties. Moreover, given the fact that NATO has to maintain a certain force-to-space ratio along the inter-German border, it cannot afford major reductions in forward-deployed forces. Any reductions in U.S. forces also entail problems. For one thing, rebasing U.S. ground forces from Europe to the United States is likely to be costly and will encourage Congress to reduce the authorized strength of the U.S. Army.

Political issues aside, NATO appears to have trouble developing a broad conceptual vision of what kind of conventional arms control regime would be beneficial to its security. Some commentators have proposed establishing common ceilings on tanks and artillery in Central Europe. Such figures as 10,000 tanks and 4,000 artillery tubes for each alliance have been discussed. This approach has the seeming advantage of sparing NATO from having to make more than nominal cuts in its forces levels. Yet, it is unclear whether a mutual pull-back from Central Europe would appreciably change the magnitude of the Soviet military threat. The reductions would not alter the Soviet Union's proximity to Europe (or NATO Center's lack of depth). Nor would they deprive the Soviet military of the inestimable advantages that accrue to the attacker. Most importantly, Moscow would retain tremendous reserves in its western military districts, which could be used rapidly to reinforce remaining Soviet forces forward deployed in Europe. This threat could be eliminated only by stripping the Soviet's western military districts of most of their reserve formations. It is inconceivable, however, that Moscow would agree to go that far.

It is, in short, difficult to conceive of a conventional arms control agreement that would both substantially reduce the Soviet threat to NATO and command Moscow's support. All that we have seen so far seems to suggest that Moscow is merely interested in exchanging its present theater force posture, with its evident emphasis on brute numbers and blitzkrieg strategies, for a more subtle, sophisticated, and flexible force arrangement which is less visibly provocative, yet retains for the Soviet leadership a full range of warfighting and war winning options in Eurasia.

18

The Role of Verification in Arms Control

Sidney N. Graybeal and Patricia Bliss McFate

Arms control agreements are a reflection and an extension of current U.S. political philosophy and military objectives. They contribute to stability by removing uncertainties and increasing understanding between countries. They neither save money nor have a positive impact upon the national economy, but their contribution to U.S. and international security makes them well worth their costs.

Arms control has been an integral part of U.S. strategic planning over the past thirty years. The ways in which it has been characterized, however, have varied from the assessment of it as a positive contributor to strategic stability and national security to the view that it is a detriment to maintaining and enhancing strategic strength. A recent swing from opposition to endorsement caught a number of observers by surprise. Having raised considerable questions about the ability of arms control negotiations to deal effectively with a country accused of striving for world domination and "cheating" on existing arms control agreements, the Reagan Administration, in its final years, is aggressively pursuing agreements with that country, as represented by the INF Treaty and the ongoing START and forthcoming conventional arms control negotiations.

Arms control can and should contribute to national security. However, in today's world, there must be assurances that all parties to an agreement are complying with its provision; thus, verification has become an essential part of the arms control process. Verification has taken a wide swing in popularity in the last eight years. It has been used to prove that agreements with the Soviet Union are not possible, nor in the interest of the United States, and more recently, to prove that agreements are both possible and desirable. Along the way, the Reagan

Administration's semantical and technological stumbling blocks to arms control have delayed new agreements and undermined existing ones. Moreover, the administration has raised verification requirements to unreasonably high levels calling for excessive verification regimes. Recent concerns about verification (some real, others manufactured) and the apparent inability to resolve real and purported noncompliance issues could jeopardize the whole arms control process.

Arms control agreements should be judged primarily on their military and political significance. However, in recent years, verification has become the central factor in evaluating the utility and effectiveness of past, present, and future arms control agreements and the whole arms control process. The meaning of the term also has varied from speaker to speaker, and its scope has expanded beyond the bounds of standard definition.

What is now referred to as the verification process goes beyond monitoring and evaluation into areas of designing verification regimes and formulating responses to noncompliance. The process includes the determination of compliance with existing agreements, policy decisions about what constitutes adequate or effective verification, the design and negotiation of regimes to meet security requirements, the implementation of verification provisions of completed agreements, and the determination of appropriate responses to ambiguous situations or clear noncompliance with specific provisions of the agreements. This view of verification as an all-encompassing process should be taken into account in assessing its role.

Verification Criteria

Consideration attention has focused on what constitutes "adequate" (the term used during the Nixon, Ford, and Carter Administrations) or "effective" (the Reagan Administration's term) verification. Although many would argue that the distinction between these terms is solely "in the eyes of the beholder," the latter term seems to place emphasis on the production of a desired effect and suggests the use of more demanding standards. In many respects, the positions taken by the Reagan Administration on verification requirements reflect this change in nomenclature. The approach to the verification of the INF Treaty and the resulting treaty provisions have produced the most extensive and rigorous regime ever successfully negotiated. In our view, it is probably more than required for effective verification if the criterion of military significance is applied in evaluating the regime.

In fact, both the Carter and Reagan Administrations have utilized this criterion. In testimony given before the Senate in January 1988, Ambassador Paul Nitze described the INF verification regime as "effective," meaning that "if the other side moves beyond the limits of the treaty in any militarily significant way, we would be able to detect such a violation in time to respond effectively and thereby to deny the other side the benefit of violation."

This is essentially the same definition used by former Secretary of Defense Harold Brown during the SALT II hearings, when he explained that "any Soviet cheating which would pose a significant military risk or affect the strategic balance would be detected by our intelligence in time to respond effectively."

The concept of "effective verification" is related to the nature and quantity of military forces considered necessary for maintaining stable deterrence and to the scope and nature of the agreement. For example, those who subscribe to a minimum deterrence theory and consider present strategic forces excessive are unlikely to see the need for rigorous verification regimes. On the other hand, those who are concerned about the ability of present strategic forces to provide an adequate deterrent will press for stringent verification provisions. The latter group clearly dominated the design of the INF verification regime and will attempt to determine the scope of the START verification provisions.

Similarly, the nature and scope of an agreement will also be factors in an assessment of the effectiveness of the verification programs. The larger the size and survivability of the remaining strategic forces, the lower will be the concern over possible cheating or circumvention. Significant reductions will cause a more careful evaluation of the effectiveness of the regime.

Because arms control agreements are essentially political instruments, any violations, real or perceived, take on major political significance. The several Reagan Administration reports to the Congress on Soviet noncompliance with arms control agreements, the press coverage associated with these reports, and the frequent references made to them by senior officials and several members of Congress attest to their political significance, regardless of their military importance. Official analyses of the potential military significance of these real and purported noncompliance issues are rare, suggesting that none of the issues meets the criterion of military significance.

This criterion is compatible with a verification regime which relies primarily on national technical means (NTM) of verification, essentially unilateral intelligence capabilities complemented by cooperative measures; but it stops short of the highly intrusive and potentially destabilizing short-notice, challenge inspections of suspect sites — the "anywhere, anytime" on-site inspections. The politically significant criterion, on the other hand, demands extensive, extremely stringent, and highly intrusive verification provisions.

Verification Problems

The question is frequently raised as to whether the "gap" between our monitoring capabilities and our verification requirements is not increasing to the detriment of future arms control agreements. The gap depends, in large measure, upon one's perception of what is required for adequate or effective verification. Increasingly complex arms control agreements and concerns over possible noncompliance are taxing our NTM and demanding increased analytical capabilities from the intelligence community. Inasmuch as future arms control agreements are going to rely on NTM as the foundation for assuring adequate verification, it is imperative that these capabilities for intelligence collection and analysis be maintained and strengthened. This is not only in our national security interests, but also will provide the necessary confidence for assuring compliance, in the militarily significant context, with the provisions of present and future arms control agreements.

Although the current INF Treaty can be effectively verified using NTM supplemented by the agreed-upon cooperative measures, including on-site inspections of declared facilities, the forthcoming, militarily significant START agreement will be considerably more complex and could strain our monitoring capabilities. As reported in *The New York Times* on February 17, 1988, members of the Senate Intelligence Committee have warned the Reagan Administration that "current surveillance systems, unless modernized, will not be adequate to monitor Soviet compliance with a proposed treaty to reduce long-range nuclear missiles." Senator David Boren (D-OK), chairman of the committee, said that the committee had requested more money than currently planned for the modernization of "technological systems," generally understood to include satellite surveillance, and he stated that "What we have now is insufficient if there were to be a START agreement." Specific intelligence budget figures are classified, but it is clear that the Senate Intelligence Committee is recommending significant additional expenditures for improving our NTM to provide the capabilities for monitoring future arms control agreements. These expenditures are worthwhile both for enhancing arms control monitoring capabilities and for improving overall intelligence capabilities which are essential to U.S. security with or without any arms control agreements.

Even with improved NTM, future arms control agreements are going to require supplemental cooperative measures of the nature and type agreed to in the INF Treaty. Such cooperative measures include detailed exchanges of data, extensive notifications of certain activities, arrangements to enhance NTM, and comprehensive, on-site inspections of *declared* facilities. There will be a valuable synergistic effect among these cooperative measures, including on-site inspections; the whole will be greater than the sum of the parts.

On-site inspections can contribute to achieving adequate or effective verification; however, they are not the panacea that many profess. They make cheating more difficult and costly, but they will not prevent a determined violator.

There are three general types of on-site inspections, each with its strengths and weaknesses:

- *Routine or short-notice inspections* of declared facilities contribute to confidence in the agreement and make cheating more difficult, as with the INF Treaty.

- *Invitational inspections* can remove ambiguities and reduce uncertainties. These should be given more attention.

- *Short-notice challenge inspections* of suspect sites are highly unlikely to reveal any violation, could be detrimental to U.S. security, and could become a basis for undermining the confidence in the agreement by the manner in which they are implemented. There is also the danger that having achieved some form of on-site inspections, they will be oversold, and thus generate a false sense of security. The disadvantages of such "anywhere, anytime" inspections exceed their advantages; it would be better to rely on our NTM, supplemented by cooperative measures and some limited routine and invitational on-site inspections for deterrence purposes.

The potential security implications of short-notice challenge, on-site inspections of declared facilities and activities merit some further attention. For example, there are often sensitive programs or activities being conducted in or near the declared facility which will be subject to short-notice inspections. The Soviet inspectors must be able to determine that there are no treaty-limited items present, but, at the same time, not acquire information on sensitive programs or activities. Such inspections can have an adverse and potentially costly impact on American industries, and raise difficult security and legal questions. If a facility must shut down during an inspection, who bears the costs? There is frequently the potential loss of proprietary information; how is this determined, and who is responsible? In addition, some facilities are also involved in very sensitive "black" programs in which security considerations are of paramount concern; they may need to remove their program from a facility subject to Soviet inspection. Who bears the cost of the move or the potential loss of such future business by that company?

Implementation of complex and detailed provisions for on-site inspections will require careful attention and considerable resources. During periods in which there are constructive and cooperative working relations between the United States and the Soviet Union, both sides want to see the arms control process advance in a mutually advantageous manner; in these periods, it is possible to implement provisions with a minimum of friction and controversy. However, the potential for mischief making by either party is inherent in the implementation of the provisions, should the climate return to a cold war environment. For example, what would happen if, intentionally or by innocent oversight or neglect, an on-site inspection team requesting a short-notice challenge inspection was not taken to the designated facility within the prescribed time or not allowed into certain parts of the facility? What if a required notification is not received on time, or not at all? What if the resident inspectors are denied certain privileges they consider inherent in their diplomatic status? There are literally hundreds of these opportunities for mischief making, which in an adversarial climate, could be blown into an expansive and controversial noncompliance record; this record, in turn, could be used to undermine the whole arms control process and jeopardize U.S.–Soviet relations.

The INF Treaty established a Special Verification Commission (SVC) to resolve compliance questions and agreed-upon measures to improve the viability and effectiveness of the Treaty. The ABM Treaty of 1972 established a Standing Consultative Commission (SCC), which was also involved in SALT II implementation and compliance issues. The functions and operations of these two bodies are essentially the same. In our view, the SCC has proven to be an effective forum for implementing bilateral strategic arms control agreements and for resolving most compliance issues; thus, we do not see the need for establishing a new, separate body for the INF Treaty. It would be more cost effective and efficient to combine the SVC and SCC into a single body responsible for handling implementation and compliance issues for the ABM Treaty, the INF Treaty, a START agreement, and any other future bilateral agreements with the Soviet Union.

Although verification was not a major issue in the ABM Treaty negotiating and ratification process, some rather serious ambiguities and compliance issues have arisen over the sixteen years' experience with the treaty. Those compliance issues with potential military significance require U.S. attention and prompt resolution. To date, the only compliance issues which meet this criterion are those associated with the Soviet ability to rapidly upgrade a portion of their extensive surface-to-air missile (SAM) system by giving it an effective ABM capability. Ambiguous Soviet activities, such as testing SAM components in an ABM mode, mobile ABM-capable radars, and rapid reloading of likely ABM launchers, fall into the potentially militarily significant category; the Krasnoyarsk radar — while a clear violation of the ABM Treaty — does not meet this criterion.

Another problem related to verification involves the concern over "breakout," which occurs when one side abrogates an arms control treaty and threatens the other side with forces in excess of treaty limits. In the case of START, Congressman Les Aspin (D-WI) and many of us believe that the real verification challenge is breakout rather than cheating. Under the current draft START agreement, the Soviet government could stockpile warheads or re-entry vehicles for ballistic missiles, stockpile air-launched cruise missiles (ACLMs) for airplanes other than heavy bombers, and stockpile test and spare intercontinental ballistic missiles (ICBMs) and sea-launched ballistic missiles (SLBMs). The START counting rules for warheads on ICBMs, SLBMs, and ALCMs on heavy bombers leave room for installing about 3,000 extra warheads over the START limit of 6,000. By combining the stockpile and counting rule breakout potentials, over 12,000 weapons could become available for breakout, a significant threat to the United States. Thus, the START verification challenge becomes one of detecting and identifying Soviet activities and intentions associated with achieving this breakout potential in sufficient time to take appropriate responses.

Resolving Verification Uncertainties

Before signing new bilateral arms control agreements, there are four steps which should be taken to deal with verification uncertainties, breakout, or potential non-compliance issues:

- Since NTM is the foundation for verification, the president of the United States should direct the intelligence community to provide an all-source assessment of its capabilities to monitor the limitations proposed in the agreement.

- The president should then direct the Department of Defense to identify specific research and development programs, with their timing and costs, which would assure continued U.S. security in the event that the Soviet government took advantage of one or more of the uncertainties in our monitoring capabilities; this information regarding "hedges" and their costs should be provided to Congress.

- Congress should recognize the verification uncertainties associated with the agreement; recognize the scope, nature, timing, and costs associated with the hedges against potential cheating; and, when providing advice and consent, accept both the uncertainties and the costs. Thus, verification uncertainties and funding of hedges or safeguards should be tied together: if either or both are unacceptable, the agreement should not be ratified.

- The executive branch and Congress should be prepared to amend, withdraw, or abrogate the agreement should the Soviet government take militarily significant action inconsistent with the agreement. This intended action should be made clear during the ratification process and promptly be implemented should such a serious situation arise.

Summary

Effective verification will continue to be essential for the attainment and viability of arms control agreements and for the maintenance of the arms control process. However, there is considerable controversy over what constitutes effective verification and the price to be paid for it. Application of the test of military significance in judging the effectiveness of verification regimes is consistent with U.S. security and will avoid unrealistic verification requirements and exhorbitant costs.

The INF Treaty contains the most extensive and rigorous verification provisions ever negotiated for any arms control agreement, and its implementation will be costly in both dollars and personnel. The provisions are probably excessive for protecting U.S. security, but necessary to obtain ratification in the current environment. In addition, they will provide a useful precedent for a more complex and militarily significant START agreement, which will contain a markedly more demanding verification regime with commensurate increases in the resources required for its implementation.

Verification has become *the* criterion for judging arms control agreements. In fact, in today's environment, the verification tail seems to be wagging the arms control dog. Some are beginning to question if the U.S. may not be generating unnecessary verification procedures and costs to the long-term detriment of arms control. Verification costs are becoming excessive in terms of what is truly required to protect U.S. security, particularly if strategic stability is assured by making our strategic forces increasingly more invulnerable. However, even with these excessive costs, the overall benefits resulting from confidence that all parties are complying with arms control agreements outweigh current and likely future verification costs. We reach this conclusion based on our belief that arms control has contributed to, and will continue to contribute to, maintaining national security, to improving relations between the United States and the Soviet Union, and to creating a stable environment marked by constructive international relations.

19

Dealing with Future Treaty Implementation and Compliance Questions

Michael Krepon and Sidney N. Graybeal

The treaty eliminating intermediate- and shorter-range nuclear missiles represents a major diplomatic achievement for several well-advertised reasons. Significant numerical imbalances favoring the Soviet Union will be negated, and many useful precedents have been set for future agreements, particularly in the area of cooperative and intrusive verification. The INF Treaty's design also creates important incentives for compliance, especially by imposing high opportunity and financial costs for cheating that provides only marginal returns.

All of this is welcome news. But the INF Treaty represents some bad news, as well: this agreement is extraordinarily complex, even by the usual arms control standards. The annual Arms Control and Disarmament Agency (ACDA) publication, *Arms Control and Disarmament Agreements*, tells the tale: the 1963 Limited Test Ban Treaty can be reproduced on two pages; the 1972 Antiballistic Missile Treaty requires nine; the unratified SALT II Treaty, signed in 1979, covers thirty-one pages. The INF Treaty text has yet to be reproduced in this format, but the U.S. Government Printing Office text of the treaty and its accompanying memorandum of understanding, protocols, annexes, photographs, and site diagrams runs to 329 pages. A strategic arms reduction treaty is likely to be even longer.

The length of these agreements is a testament to the growing complexity of the issues they cover. Arms reduction is more consequential and difficult than arms control, just as limitations on mobile and cruise missiles are far more demanding than counting missile silos and submarine launchers. The trend toward complexity in treaty texts is mirrored by more intensive congressional involvement in their review. The Senate's deliberations before providing consent to ratification of the Limited Test Ban and ABM Treaties fit within a single volume of hearings. The contentious deliberations over SALT II produced twelve volumes of testimony and debate; even the relatively smooth passage of the INF Treaty required eleven volumes.

The implementation of these agreements has also grown far more complex. Long gone are the days of the Nonproliferation Treaty ratification hearings, when the Senate agreed to leave implementation questions and safeguards in the hands of an autonomous agency of the United Nations. In SALT, implementation questions were entrusted to a bilateral Standing Consultative Commission (SCC), with the relevant congressional committees periodically advised of the results of these secret deliberations. In INF, the Senate became a direct participant in implementation matters because many of these details were included in the treaty text.

In addition, the bureaucratic apparatus required to address treaty implementation and compliance questions has grown considerably. Three new government offices have been created by the INF Treaty: an on-site inspection agency affiliated with the Pentagon, an office in ACDA to handle compliance questions, and an INF monitoring office in the Central Intelligence Agency (CIA). A recently established office within the State Department, the Nuclear Risk Reduction Center, will handle the message traffic for treaty implementation. The creation of new bureaucracies also creates new gray areas of divided responsibility. Proper oversight and coordination within the executive branch will be more difficult than even before.

The geometric growth in complexity, Congressional involvement, and bureaucracy does not bode well for those who approve of arms reductions but also believe in Murphy's Law. These trends create new possibilities for inadvertent problems and missed communication. All can easily become evident in new superpower accords; they can become far worse in future multilateral agreements. Significant problems in treaty implementation and compliance can still be avoided as long as U.S. and Soviet political leaders are committed to succeed and have the support to do so. Nevertheless, the sheer level of complexity involved suggests multiple opportunities for mischief making and crossed wires. At the very least, high-level political officials can expect to spend more of their time than seems warranted on extremely obscure treaty implementation and compliance questions. More likely, small problems can accumulate, receiving inordinate prominence and requiring scarce political capital to resolve.

If this complex new era of bilateral arms reductions is to succeed, intense cooperation will be required on three fronts. First, the superpowers must cooperate to an unprecedented degree. If concerns over each other's strategic activities and intentions grow over time, the process of arms reduction will be short-circuited. Second, the executive and legislative branches must develop and sustain a more cooperative working relationship. Finally, intense cooperation and coordination will be required within the executive branch to ensure that treaty implementation and compliance questions are handled expeditiously and properly.

Much attention has been focused on how to improve the first two areas of cooperation, and hardly any on the third. This is natural, since Americans focus their attention on the contentious negotiations leading up to an agreement and on the battle over ratification. But the process of treaty implementation is also critical: if done poorly, compliance problems mount, superpower relations sour, and the negotiating process stalls. This essay addresses the third area of cooperation needed to succeed at arms reductions: the establishment of an effective U.S. management system for handling treaty implementation and compliance questions.

The bureaucratic arrangements chosen matter only at the margins. Far more important are the policy guidelines that bureaucrats operate under and the orientation that government officials apply to their tasks. Good civil servants can be hamstrung by poor policies; they can also succeed in government despite a cumbersome bureaucracy, just as poorly qualified or motivated individuals can complicate even the most streamlined design. Moreover, some compliance problems are quite beyond our control: if the Kremlin chooses to badly bend or break the rules established by arms control agreements, it will remain free to do so regardless of what organizational approach the executive branch adopts. This was the case between 1979 and 1983, when the Soviets heavily encrypted telemetry, began flight-testing two new missiles instead of the one "new type" permitted under the unratified SALT II Treaty, and initiated construction of the Krasnoyarsk radar, a clear violation of the ABM Treaty.

Even at the margin, however, these issues can have considerable political salience. Proper internal oversight might prevent some minor compliance questions from arising. The choice of bureaucratic mechanics for handling disputes can also simplify their successful resolution, or magnify their significance. This is especially true for treaties as complex as the INF accord. The procedures it invokes are so new that problems can readily arise during the implementation period. Nor can the treaty text and accompanying documents possibly address every conceivable contingency. After all, this is the first time that U.S. and Soviet negotiators have attempted to work out such extensive and intrusive verification arrangements; circumstances are bound to arise during the implementation process that raise compliance questions.

Given the contentiousness and significance of the subject matter, it would be wise to establish institutional arrangements for implementation and compliance that facilitate prompt, objective evaluation of Soviet actions, helping U.S. decision-makers to distinguish between militarily significant breaches and modest, perhaps inadvertent, problems. In either case, bureaucratic arrangements should maximize the likelihood that problems are resolved as successfully and expeditiously as possible. Whatever bureaucratic design is chosen for handling treaty implementation and compliance questions must therefore meet three critically important criteria: institutional arrangements must be (1) conducive to prompt but well-coordinated evaluation, (2) resistant to manipulation, and (3) disposed to problem-solving approaches.

Older bureaucratic models do not fit the new level of complexity reflected by the INF Treaty. For the SALT accords, a single institution was established to handle implementation as well as compliance questions — the SCC. These tasks were not considered to be time-consuming. An extremely small office was established in ACDA for the U.S. Commissioner and his Executive Secretary. The SCC would regularly meet only twice a year, for perhaps two months at a time, and occasionally in a special session called by one of the parties.

There are several reasons why the old SCC mode of operation is inappropriate for the INF Treaty. To begin with, unlike prior agreements, the INF accord includes basic implementation instructions. These instructions mandate hundreds of inspections and the establishment of a large cadre of inspectors qualified to carry them out. The INF Treaty also provides for the transmission of significant amounts of data on a continuing basis. In contrast, the SALT accords required only modest data exchanges. Moreover, SALT did not expressly permit inspections, and none were carried out under the SCC's charter. The choice of institutional arrangements has been further complicated by the bitterly hostile view of some administration officials, members of Congress, and outside observers to the work of the SCC. Apparently at no time in the course of the INF negotiations did the U.S. side propose a formal mechanism like the SCC to handle compliance disputes. The Soviet Union reportedly insisted on such a mechanism.

In the end, both sides agreed to handle treaty implementation and compliance functions separately. For implementation, both established their own On-Site Inspection Agencies. For compliance matters, they agreed to establish a Special Verification Commission (SVC). At the same time, the treaty specifies that each side's Nuclear Risk Reduction Centers (NRRC), established in September 1987, would become the channel used to exchange data, notifications, and to "provide and receive requests for cooperative measures."

In the United States these new institutions have been superimposed on existing bureaucratic flow charts for handling treaty implementation and compliance questions. In 1982, the Reagan Administration established a high-level Arms Control Verification Committee (ACVC) chaired by the president's national security adviser. Its charter is to assess the verifiability of arms control proposals, review monitoring requirements, and analyze Soviet compliance with existing agreements. In practice, the ACVC has focused almost exclusively on the third task.

The work of the ACVC has been carried out by an analysis group and a policy group, with common core membership at the assistant secretary or deputy assistant level. The analysis group is co-chaired by the director of ACDA's Verification Bureau and the chief of the Arms Control Intelligence Staff at the CIA. Its primary function is to correlate Soviet military activities with the Kremlin's obligations under arms control accords. The analysis group is not supposed to draw conclusions about Soviet noncompliance; this is the responsibility of the ACVC policy group, co-chaired by the Department of State's Assistant Secretary for Politico-Military Affairs and the Pentagon's Assistant Secretary for International Security Policy. Representatives from the Department of Energy, Joint Chiefs of Staff, and the Defense Intelligence Agency also participate in these deliberations. Interagency working groups (usually at the office director-level provide support for the ACVC analysis and policy groups.

The necessity for two separate groups with overlapping membership is largely due to a longstanding mutual desire by intelligence community professionals and policy makers to maintain the distinction between analysis of Soviet military activities and conclusions about noncompliance. The decentralized bureaucratic apparatus for doing so, however, is a Reagan Administration phenomenon. In previous administrations, analysis was carried out by the intelligence community without ACDA participation, and compliance assessments were directed by the National Security Council staff instead of by an interagency committee co-chaired by State and Defense.

The decentralized approach favored by the Reagan Administration reflects this particular president's managerial references. For most of the administration's tenure, the awkwardness of these arrangements has been quite evident, especially with strong Cabinet-level disagreements over arms control and U.S.–Soviet relations. To streamline the process and to deal with new requirements associated with the INF Treaty, the Reagan Administration has established an INF implementation steering group. This interagency group, unlike previous subcabinet-level bodies, is chaired by an NSC staffer. Its office director level working group is chaired by an ACDA official. Meanwhile, the ACVC analysis and policy groups, which have general jurisdiction over treaty implementation and compliance questions, continue to function.

Reagan Administration officials acknowledge that the resulting organization chart has a Rube Goldberg-like appearance, since difference interagency groups exist with comparable mandates chaired by different parts of the national security bureaucracy. For example, the reports prepared by On-Site Inspection Agency (OSIA) teams are being sent to the intelligence community's monitoring office, the ACVC analysis group and the INF implementation steering group. Concerns are automatically raised about the division of labor after OSIA reports are filed and the potential for differing interpretations of the data arises. Separate channels are also growing for handling treaty compliance matters with the Soviet Union. In addition to the new U.S. representative to the SVC and the existing American Ambassador to the SCC, the Reagan Administration is planning to establish a separate institution to handle START compliance questions.

Administration officials downplay the potential for crossed wires and missteps because of this organizational complexity. They point out that the same individuals are working on these problems, even if the channels for handling them have proliferated. They also note that divisions of opinion regarding the value of nuclear arms reduction agreements with the Soviet Union have dramatically faded within the executive branch, and there is currently a strong desire to make implementation of the INF Treaty a success in both Moscow and Washington.

These arguments have merit, but they do not inspire great confidence over time. While an impressive consensus has grown in support of the INF Treaty, contentious debates continue over the value of START, a comprehensive test ban, and conventional arms reductions. Nor is it likely that policymakers engaged in future negotiations will be completely harmonious in their perspectives on Soviet treaty implementation and compliance. It would also be foolish to predict that the current patch of good superpower relations will last throughout the thirteen year implementation period of the INF Treaty. If U.S.–Soviet relations deteriorate, American policy makers will have enough substantive problems to address without adding to them relatively modest, but politically inflated, concerns over treaty implementation and compliance.

For all of these reasons, the existing system for handling treaty verification and compliance questions cries out for simplification and restructuring. Current arrangements are far too confusing, while inviting mischief making and the generation of unintended problems. These arrangements are also too complex and redundant to provide prompt, but well-coordinated, evaluation of intelligence, while being resistant to manipulation and disposed to problem-solving approaches — our three basic management objectives for successful treaty implementation and compliance. What follows is a list of suggestions on how these objectives might be achieved more easily, keyed to the new institutions created by the INF Treaty but applicable to future bilateral agreements, as well.

On-Site Inspection Agency (OSIA)

The Reagan Administration was divided over the bureaucratic placement of the OSIA, with some counselling a location in ACDA and others pushing for the Pentagon. Largely for budgetary, logistical, and manpower reasons, a Pentagon connection was chosen, even though ACDA has more of a statutory authority to carry out this function. Faced with near-term requirements to carry out and administer the large number of inspections permitted under the INF Treaty, the administration chose bureaucratic arrangements likely to carry out this mission with the least friction and the most available resources. OSIA's first year budget of $89 million was covered by reprogramming funds from other Defense Department accounts. ACDA, entering its twenty-sixth year of operation, only has a budget of slightly more than $30 million. Full-time personnel assigned to OSIA number 150, while ACDA has almost 200 professionals; when detailees from other agencies are added to these numbers, OSIA doubles in size and becomes larger than ACDA.

ACDA clearly would have had difficulties absorbing the OSIA, given the relative sizes and budgets of the two agencies. Additional problems are created by the different missions of the two agencies. OSIA has a straight-forward operational mission: to carry out inspections and to escort Soviet inspectors. To assist OSIA in these tasks and to help insulate it from the vicissitudes of political debate, it is specifically enjoined from venturing into the policy arena. In contrast, ACDA has a mandate to become deeply involved in policy questions. Due to its small size, a change in relatively few officials at the top can alter dramatically ACDA's bureaucratic role, from championing new agreements to disparaging their worth. As a result, ACDA has become a lightning rod for criticism. As long as this remains the case, it is best to keep the OSIA separate from it. If, on the other hand, ACDA becomes depoliticized over time by changing its orientation from policy to research and operational concerns, a future merger with the OSIA could make sense.

While solving immediate logistical, budget, and manpower problems, the OSIA's placement in the Pentagon raises several others. An immediate problem is how to provide for proper coordination among other agencies who have a direct stake in how the OSIA carries out its implementation tasks. The Reagan Administration has addressed this problem by issuing an internal directive mandating direct involvement by ACDA, the Department of State, and the Federal Bureau of Investigation (FBI) in the OSIA hierarchy. Under this directive, an official from ACDA is the OSIA Director's principal deputy, while the State Department and FBI provide deputy directors. FBI liaison is critical for hosting Soviet inspection teams, as it has domestic responsibility for counterintelligence operations. Liaison with U.S. military operations is established through the Directorate for Plans and Policy (J-5) of the Joint Chiefs of Staff. To reinforce the operational, nonpolitical nature of the OSIA mandate, a highly professional and skilled military officer with two tours at the American Embassy in Moscow and previous assignment as Chief of the U.S. Military Liaison Mission in East Germany, Brigadier General Roland Lajoie, was chosen to be the first director of the new agency.

The Reagan Administration's approach to coordinate OSIA activities within the executive branch makes good sense, but larger questions remain about the OSIA Director's chain of command. Since the SALT agreements were signed in 1972, the Pentagon's chief acquisition officer or scientist has had responsibility for making sure that U.S. military activities are in conformity with treaty obligations. As a result, the OSIA Director is to report to the Under Secretary of Defense for Acquisition, who reports to the Secretary of Defense. The Under Secretary for Acquisition is chairman of an executive committee inside the Pentagon consisting of the Under Secretary of Defense for Policy and Chairman of the Joint Chiefs of Staff or their delegates. These arrangements ultimately provide the Secretary of Defense considerable influence over OSIA operations.

This could raise problems if and when a future Secretary of Defense strongly wishes to use the OSIA to demonstrate displeasure with Soviet activities or does not believe continued implementation of an agreement to be in the U.S. national interest. The operational activities of the OSIA should not be influenced by a particular policy bias. This overriding concern, which helped swing the scales against the placement of the OSIA within ACDA, should also apply to the Pentagon.

These concerns over the improper utilization of executive branch machinery for handling treaty implementation and compliance matters are not merely hypothetical. In its November 1987 report, "Intelligence Support to Arms Control," the House Intelligence Committee found that the SCC had been "under-utilized in resolving compliance problems" during the Reagan Administration, and that Pentagon officials most critical of the SCC made "no concrete initiatives" to improve its operations. A more direct parallel to the OSIA is the channel for U.S. and Soviet naval officers to discuss dangerous incidents at sea created by a 1972 superpower agreement. In June 1985, an intervention by Secretary of Defense Caspar Weinberger to protest the shooting of Major Arthur D. Nicholson led to the cancellation of this annual meeting between naval officers.

Current arrangements where the Director of OSIA, a senior active duty military officer, reports to the Under Secretary of Defense for Acquisition and to the Secretary of Defense require review. Ideally, the OSIA should be truly autonomous, with its own budget and without being dependent upon other agencies for manpower and the Pentagon for budgetary and logistic support. Resources are probably too scarce for this to happen, at least in the short run. Nevertheless, more autonomy for the OSIA is needed to provide assurance that its director is not too dependent upon the bias of a single institution. To minimize this problem and to improve coordination with related treaty compliance activities, a better alternative would be for the OSIA Director to report to the president's national security adviser. This approach provides no guarantees against politicization or mischief making, but it is more in character with the interagency nature of the OSIA's tasks.

Special Verification Commission (SVC)

During the INF Treaty ratification debate, Reagan Administration officials attempted to point out the differences between the SVC and its predecessor, SALT's Standing Consultative Commission. Unlike the SCC, the new INF consultative body would meet on an ad hoc basis at varying levels of representation, and neither side would be permitted to decline to attend meetings called on short notice. The agreed mandate of the SVC was also much shorter than the SCC. In the ABM Treaty, the SCC is empowered to consider compliance questions, exchange information, deal with interference with national technical means of verification, "consider possible changes in the strategic situation which have a bearing" on treaty provisions, agree upon procedures for dismantling or destroying weapons, consider proposals for increasing the viability of the treaty, and even consider further measures limiting strategic arms. In contrast, the INF Treaty calls on the SVC "to promote the [treaty's] objectives and implementation" in two ways: to "resolve questions relating to compliance," and to "agree upon such measures as may be necessary to improve the viability and effectiveness of the treaty."

In actuality, there appears to be little difference between the mandates of the SCC and the SVC. The SCC has the latitude to call special sessions on short notice, although this mechanism became a casualty of the politicization of compliance diplomacy during the Reagan Administration. (Once, the administration called for a special session a few weeks before one of the SCC's semiannual meetings. When the Kremlin declined to attend because of this, its refusal was noted by administration officials as evidence that the Soviets were misusing channels for resolving compliance problems.) The SCC also has the ability to deal with compliance questions at varying levels of authority through the practice of establishing working groups. Even the language of the INF Treaty establishing the SVC, while more succinct than that of the SCC, is broad enough to provide for comparable latitude.

In the final analysis, the perceived requirement to establish separate consultative mechanisms for SALT and INF was driven exclusively by political considerations. To some Reagan Administration officials and their colleagues on Capitol Hill, the SCC is a symbol of all that is wrong with arms control agreements and with the handling of U.S.–Soviet relations by prior administrations. Their concerted effort to denigrate the worth of the SCC overlooked previous successes in resolving compliance problems, as well as the fundamental reality that this institution can be no more effective than the instructions under which its commissioners operate. Nevertheless, critics succeeded in tainting the SCC to the point where no cabinet-level official argued that it should be given expanded authority to deal with INF compliance questions. Even those who did not denigrate the SCC were quite satisfied to establish a separate channel for INF compliance diplomacy in order to smooth the new treaty's implementation process.

Sensitivity over the SCC/SVC parallel during the INF Treaty ratification debate extended to the point where administration officials declined to establish an office and to designate a staff within the executive branch for SVC activities. Ostensibly, the reason for this decision was to maximize flexibility — a rationale that few found persuasive. After the treaty entered into force, the administration established an office in ACDA for the SVC, and selected a U.S. representative and his deputy. This office will operate alongside the ACDA office for the SCC. Neither will have its own professional staff; instead, they will draw on officials working elsewhere in the executive branch. If a START agreement enters into force, a third institution will be created, the Joint Compliance and Inspection Commission, which presumably will require yet another office.

In the future, American interests would be better served by consolidating these activities and by dedicating professional expertise to work on compliance problems. Effective compliance diplomacy requires special skills and institutional memory that can best be developed in a single office staffed by seasoned professionals. Separate offices increase the likelihood for crossed wires and opportunities for Soviet manipulation. Two other advantages can accrue from establishing a central office to handle compliance questions: first, a permanent office staffed by competent professionals can establish effective working relationships with their Soviet counterparts. This can provide the U.S. government with important insights to Soviet behavior and suggestions for appropriate American responses. Second, this expertise can also be helpful to U.S. negotiators seeking to foreclose potential avenues of Soviet mischief making in new accords. The Kremlin has always maintained a close liaison between its negotiating team and its team handling compliance questions. In this instance, we would do well to follow the Soviet lead.

One concern raised by the establishment of a consolidated office to handle compliance matters is that U.S. officials will develop a vested interest in smoothing over Soviet misbehavior. The possibility of Soviet manipulation exists whether or not the United States assigns professional expertise to an office to deal with these problems. On balance, however, there are clear advantages in establishing a single compliance-resolution office within the executive branch. Ad hoc responses to compliance problems can provide new opportunities for the Kremlin to test the United States and to exploit inconsistent responses. Given the controversy surrounding the work of the SCC during President Reagan's watch, the creation of a new full-time office for compliance-resolution could also send important messages regarding a new administration's seriousness of purpose in resolving disputes that arise.

This office could continue to be placed in ACDA; or it could be moved to the executive office of the president, where the National Security Council staff, Office of Management and Budget, Office of Science and Technology Policy, and Office of the U.S. Trade Representative are currently located. The bureaucratic relocation of the office from ACDA to the executive office of the president can provide an important symbol of White House oversight and interest in the successful resolution of compliance questions. This move could also provide the office with greater budgetary independence from existing national security bureaucracies. Wherever the compliance resolution activities are located, it makes sense to combine the SCC and SVC functions into a single office, and to place additional responsibilities under START here, as well.

Intelligence Community Monitoring

The line between the intelligence community's monitoring function and the policy community's verification function has never been entirely distinct. Monitoring activities lead naturally to judgments on compliance, as is evident by disputes within the intelligence community over various citations of Soviet cheating. Nevertheless, the separation of monitoring activities from compliance judgments is a worthy goal. It is logical to place the main responsibility for treaty interpretation in the hands of the policy agencies with negotiating responsibility, while the intelligence community focuses on one of its primary responsibilities — monitoring Soviet military activities and intentions. This separation also constitutes an important restraint on the politicization of intelligence. In this context, politicization means the misuse of intelligence to support any particular bias — either that the Soviets habitually cheat or that they are disinclined to do so.

Of course, professional intelligence analysts have no way to guarantee the proper use of their product by policymakers or by the president. Political officials, however, incur penalties when they overstate or understate Soviet noncompliance. When available intelligence is misrepresented either way, political controversies erupt, American officials lose credibility, and U.S. motives are impugned. These political and diplomatic penalties are likely to grow along with the distance between available intelligence and its public misrepresentation. It is therefore healthy to try to maximize the degree to which the intelligence community's monitoring function can be insulated from the policy preferences of political appointees and the political issues associated with compliance diplomacy.

In recent years, the separation between treaty monitoring and verification has become less distinct, in part as a result of novel bureaucratic arrangements adopted by the Reagan Administration. In prior administrations, the intelligence community's analysis of Soviet behavior was coordinated by the CIA. During the Reagan Administration, analysis of Soviet activities relevant to their treaty obligations has been co-directed by the CIA and ACDA. At the same time, ACDA has remained active in the policy arena where judgments on Soviet compliance practices are made. This dual role has blurred the distinction between treaty monitoring and verification. It has not been healthy — either for ACDA or for the quality of the administration's annual reports on Soviet noncompliance.

The more deeply ACDA delves into the analysis of Soviet compliance, the more politicized administration assessments will become as long as the agency is subject to dramatic swings based on election results. This will be the case whether the stimulus for ACDA's involvement is based on support or criticism of arms control efforts. Those who are pleased with ACDA's greater profile during the Reagan Administration might therefore contemplate the consequences of a similar role in a future administration not of their liking. And those who generally wish ACDA to become a more powerful agency to support arms control efforts must ask what institutional benefits derive in the long run by taking a lead role in compliance assessments. In the future, it makes sense to curtail drastically ACDA's role in the analysis of Soviet compliance practices, with the CIA again directing these assessments. If the CIA resumes the task of coordinating assessment of Soviet military activities as they relate to treaty compliance, there is no need for the continued existence of the Reagan Administration's ACVC analysis group.

The bureaucratic arrangements chosen by the Reagan Administration to reach conclusions on Soviet compliance have not served the nation well. Given the president's tendency to split the difference between strongly held but conflicting advice, the ACVC policy group, co-chaired by the Departments of State and Defense, produced reports of Soviet noncompliance of very uneven quality. These reports have mixed citations where the evidence is strong with those where supporting evidence is weak. Problem areas with troubling as well as insignificant military implications have been largely undifferentiated. Moreover, American credibility has suffered by the creation of a double standard in compliance reporting, something the Reagan Administration has warned against strongly. For example, prospective Soviet violations are cited, while prospective U.S. noncompliance with the ABM Treaty due to the Strategic Defense Initiative are dismissed. U.S. credibility has also suffered by the well-founded perception abroad that more effort has been placed in restating compliance concerns than in resolving them in diplomatic channels. With savvy new leadership in the Kremlin, the Soviets have made deep inroads in Western public opinion on these issues. Sadly, opinion polls in Western Europe now suggest that our own allies believe the Soviet Union is likely to be a more trustworthy party than the United States for future agreements.

These polls reflect poorly on the tactics chosen by some Reagan Administration hard-liners. While a different organizational approach is unlikely to have resulted in the solution of outstanding compliance problems, they were exacerbated by the decentralized bureaucratic design chosen by Reagan Administration officials. A better alternative is offered by the creation of an interagency steering group for the INF Treaty chaired by the NSC staff. This body necessarily has overlapping jurisdiction with the ACVC policy group. Policy coordination on compliance judgments and the quality of executive branch reports will likely improve if the awkward ACVC policy group co-chaired by the State and Defense Departments is disbanded. Instead, the NSC staff-chaired intergency committee on the INF Treaty should take its place, with a broader mandate to cover other compliance questions, as well.

Conclusion

The existence of separate organizations to work on INF Treaty notifications, inspections, monitoring, and compliance resolution requires special coordination and oversight — particularly since all four are located in different parts of the executive branch. Even so, interagency disputes will inevitably continue on treaty implementation and compliance problems. Since bureaucratic impasses only serve to heighten outstanding issues, techniques must be found to resolve them as effectively and as quickly as possible, or to take differences to the president for prompt resolution. All of these offices need to be working from common guidelines.

Effective management of a verification and compliance regime requires that all four INF Treaty offices report to the same central authority. Given the different agencies involved, that central authority can only be the NSC. The logical channel for that authority is the Assistant to the President for National Security Affairs, and the person to whom he or she delegates this authority.

The potential for abuse clearly exists under an overzealous national security adviser or staff, just as it exists in a decentralized form of "cabinet government." In the final analysis, complex verification and compliance issues are particularly ill-suited for a decentralized approach to policymaking. The natural outcome is likely to be heightened disputes over politically sensitive issues, prolonged impasses, and poor coordination of technical analyses and policy judgments. Enduring disputes over treaty implementation and compliance questions also means that these volatile issues are the last to be resolved in a politically charged negotiating endgame.

The streamlined process described here should be more conducive to prompt, well-coordinated assessments, resistant to manipulation, and disposed to problem-solving approaches than the current decentralized system. New problems can easily be created, however, if an all-powerful, high-profile individual is selected to serve this coordinating function on the NSC staff. There is enough built-in friction between the national security adviser, the Secretary of State, and the Secretary of Defense without adding an "arms control czar" to this mix.

20

Mobile ICBM, ALCM, and SLCM Verification

Jim Blackwell

Mobile intercontinental ballistic missiles (ICBMs), air-launched cruise missiles (ALCMs), and sea-launched cruise missiles (SLCMs) are the three principal obstacles to a Strategic Arms Reduction (START) agreement. While the Soviets continue to oppose the Strategic Defense Initiative (SDI), it is doubtful that SDI would block an otherwise acceptable START agreement. Their less-vocal opposition to SDI over the last few months may indicate that the Soviets are relying on the next administration or Congress to remove this obstacle. Mobile ICBMs, ALCMs, and SLCMs present a common problem: they cannot be verified with high confidence by even the most intrusive verification regime; the sides will not ban them; nevertheless, they must be constrained in some way by a START agreement.

Even with a highly intrusive verification regime, there will be considerable uncertainty in verifying compliance with limits on mobile ICBMs because of their mobility and abundant concealment opportunities. This uncertainty is particularly troublesome because of the refire and breakout threats posed by covert mobile ICBMs and launchers. Nevertheless, because the Soviets are very unlikely to agree to destroy two of their newest strategic missile systems (the road-mobile SS–25 and rail-mobile SS–24), if a START agreement is to be achieved, verification regimes must be designed that reduce monitoring uncertainties as much as possible.

The primary obstacles to agreement on ALCM limits have been over how many ALCMs are to be attributed to heavy bombers (a counting rule) and how to distinguish ALCM–carrying heavy bombers from heavy bombers that do not carry

ALCMs. The Soviets prefer an ALCM counting rule that is based on the maximum loading capacity — which is the most verifiable rule. The U.S. argues that this would significantly overcount actual bomber loadings, and that attributing 10 ALCMs per heavy bomber represents a more accurate average of operational ALCM loadings. The sides have apparently overcome the long standing problem of distinguishing ALCM carriers from non–ALCM carriers, which, in some cases, involve the same aircraft model (e.g., B–52G and B–52H), by requiring separate basing arrangements. A requirement to distinguish nuclear-armed from conventionally armed ALCMs has recently surfaced, presumably due to U.S. interest in protecting the right to deploy conventionally armed ALCMs outside the treaty limits.

The two sides have agreed to seek a limit on long-range, nuclear SLCMs; however, they have been unable to find an acceptable way of verifying such a limit. Thus far, no effective means has been found to verify that conventionally armed SLCMs are not nuclear-armed. Furthermore, the United States is determined to deploy long-range, conventionally armed SLCMs, which rules out limits on all long-range SLCMs. The problem is further complicated by the fact that U.S. nuclear-armed SLCMs are deployed on surface ships and submarines which have other nuclear and conventional weapons. Since verification of limits on long-range, nuclear-armed SLCMs appears to be an intractable problem, compliance judgments may have to be based solely on intelligence estimates.

This paper examines these verification problems and provides an overview of their negotiating status as reflected in open source documents.

Mobile ICBMs

The Soviet Union has already deployed approximately 100 road-mobile SS–25 ICBMs and began deploying the rail-mobile SS–24 ICBM in early 1988.[1] Unless constrained by a START agreement, by the mid-1990s, approximately 18 to 20 percent of the Soviets' strategic warheads are projected to be deployed on mobile ICBMs.[2]

In January 1987, the United States initiated full-scale development of the road-mobile small ICBM and the rail-mobile M–X missile as a hedge against the possibility that mobile ICBMs are permitted in START or that a START agreement is not reached.[3] Nevertheless, there is considerable uncertainty over which type of mobile system will be deployed, with the administration and the Senate favoring the rail-mobile M–X and the House supporting the road-mobile small ICBM. Democratic presidential candidate Dukakis' announced intention to cancel both mobile programs further clouds the picture. Even with a Bush Administration, options may be severely limited because of tighter constraints on the defense budget.

Thus far, the United States has proposed that mobile ICBMs be banned in START because of the difficulty in verifying the number of missiles. The U.S. has agreed, however, to consider whether or not to permit a limited number of mobile

ICBMs, if an effective verification regime can be developed. Furthermore, the U.S. has agreed to work with the Soviets on the mobile ICBM verification problem.[4]

The refire threat posed by the more survivable mobile launchers suggests that the number of spare (nondeployed) missiles should be tightly constrained, and that the nondeployed missiles should not be permitted in the vicinity of deployed or nondeployed launchers. It also underlines the importance of being able to verify that there are no illegal missiles within range of mobile launchers. For both road- and rail-mobiles, compliance with limits on nondeployed missiles will more than likely be monitored by the same verification regime that the U.S. has proposed for nondeployed, silo-launched ICBMs and SLBMs.

If the same verification regime is adopted, the number of nondeployed missiles at each missile support facility (production, repair, storage and training facility, and test range) would be declared and verified initially by a baseline inspection, and by short-notice inspections thereafter. Movement between declared locations, including road-mobile operating bases and rail-mobile garrisons, would be notified. The notifications would be used to maintain an accurate missile inventory at each location, to remove a plausible explanation for an illegal missile that was detected outside declared facilities, and to help resolve disputes involving missiles that are detected outside declared facilities. Short-notice inspections of undeclared locations (or suspect sites) would be used when covert production, storage, or repair was suspected. Continuous perimeter-portal monitoring (inspectors monitoring a suite of sensors) at missile production facilities would be used to count new missiles as they enter the inventory.[5] Over time, this should improve the U.S. confidence in the total missile inventory as older missiles systems are retired.

In contrast to silo launchers and SLBM launchers, covert mobile ICBM launchers pose the same concerns as covert missiles — perhaps even more concern since the covert launchers could be used to fire "legal" nondeployed missiles. Presumably, the verification regime that would be used for nondeployed missiles would therefore be applied to nondeployed mobile launchers. If this happens, the most effective perimeter-portal monitoring location for mobile launchers would be the final assembly facility where the erector-launcher is mounted on the road transporter or the rail car.

In addition to the refire threat posed by covert nondeployed missiles and launchers, the United States is also concerned about the use of mobile ICBMs in a sudden Soviet breakout from a START agreement. Breakout, which would result in treaty abrogation as a precursor to war, could involve covert missile units marrying up with legal nondeployed missiles and launchers, or illegal nondeployed missiles and launchers, or covertly deployed units with their own missiles and launchers.[6]

The breakout threat posed by permitted nondeployed missiles and launchers can be constrained to the extent that the U.S. can establish ceilings at the lowest possible levels consistent with U.S. operational requirements and procurement practices. The Defense Policy Panel of the House Armed Services Committee, in its May 1988 report, expressed serious concerns about the breakout threat posed by legal nondeployed weapons. The verification regime for nondeployed missiles and launchers seeks to minimize the opportunities for illegal nondeployed missiles and launchers that could be used for breakout or as refires. To guard against the breakout (and refire) threat posed by operational missile units, the U.S. must be able to confirm that declared units do not have access to undeclared missiles or launchers and that there are no covert missile units. While the verification requirements for deployed road- and rail-mobiles are similar in some ways, there are unique requirements that must be considered when designing their verification regimes.

Deployed Road-Mobiles

The fundamental verification challenge posed by road-mobiles is finding a means of ensuring that missile units are not covertly deployed in the vast expanse of the Soviet Union. Even though the scope of the monitoring problem is reduced somewhat by terrain that cannot be used by mobile ICBMs, enhanced monitoring capabilities and coverage will be essential to improve the U.S. ability to detect covert road-mobiles.

At a minimum, covert regimental-sized units would be deployed in nonstandard operating bases to minimize the risk of detection by NTM. Field training would require prolonged overcast conditions and communications that could not be detected by NTM. Such constraints could impact unit effectiveness. Covert missile units could also remain concealed in buildings, caves, or under other types of cover; however, unit effectiveness would be questionable if the missile unit was unable to operate as a unit and train in field conditions. Collocation with other military units is also possible, but the covert missile unit would encounter similar unit effectiveness problems and would probably be subject to an increased risk of detection. With current National Technical Means (NTM) capabilities and coverage, there is no apparent way of fully resolving such uncertainties; thus, if road-mobiles are to be permitted, the task is to bound the problem. Congressional bipartisan support to modernize and upgrade the U.S. NTM capability to improve the U.S. ability to monitor the provisions of a START agreement could reduce some of this uncertainty in a few years.

During the Moscow Summit, the U.S. put forth some ideas on both road- and rail-mobile verification. The Soviets reportedly agreed to many of the U.S. proposals, while disagreeing or reserving on others.[7] They agreed to confine deployed missiles and launchers to relatively small restricted areas, which would be specified in the Memorandum of Understanding (MOU) on Data. Thus, any missile or launcher that is detected outside a restricted area and that is not in notified transit would be a violation. Whereas this monitoring task would still be somewhat akin to looking for a "needle in a haystack," it is certainly preferable to permitting road-mobiles to freely operate anywhere in the Soviet Union. The sides disagreed on the size of the restricted area, with the U.S. proposing approximately 25 square kilometers and the Soviets calling for approximately 100 square kilometers. Nevertheless, even 100 square kilometers would be a dramatic reduction from the size of the designated deployment areas for the Soviet SS–20 and the U.S. Pershing II and GLCM in the INF Treaty.

The sides also agreed to a number of provisions that would enhance the effectiveness of the restricted area concept without unduly constraining normal operating practices or compromising the survivability of the road-mobiles. For example, with prior notification, a limited number of missiles and launchers would be permitted to leave the area for training, maintenance, and testing.

Missiles and launchers would also be permitted to leave their restricted area for exercise dispersals and operational dispersals. Exercise dispersals would be limited in duration and frequency, and would be notified within an agreed-upon number of hours after commencement. Operational dispersals would be for national security purposes (e.g., crisis situations), and would therefore not be limited in duration or frequency. The value of the operational dispersal provision is that survivability could be increased during a crisis without violating the treaty. Upon return of the dispersed force to a restricted area, the other side would have the right to confirm the return by designating a percentage of the total road-mobile force for open display to NTM or for a post-dispersal, on-site inspection. Obviously, frequent and prolonged operational dispersals would seriously degrade the U.S. monitoring capabilities, would undermine the restricted deployment area concept, and would likely be construed as a breach of the intent of the treaty.

The sides also agreed to limit the number of launcher-unique structures (e.g., SS–25 single-bay garages with sliding roofs) to no more than the number of declared missiles on launchers in the restricted area. Finally, a specified number of times each year, each side could request that measures be carried out to enhance NTM observation of the number of treaty-limited items in the restricted area. Presumably, this would involve some type of open display at missile operating bases while demonstrating that extra launchers are not concealed in launcher garages.

These are useful steps; however, much work remains. Some of the remaining issues include: "tags" for accountable missile stages and mobile launchers to help distinguish legal from illegal ones; whether to ban liquid-fueled mobile ICBMs to reduce covert production concerns; whether all spare launchers should be located at least an agreed distance from restricted areas and non-deployed road-mobile ICBMs to reduce access to refire missiles; whether to ban missile resupply vehicles for mobile ICBMs to also reduce the refire threat; and criteria to be used in determining what types of undeclared locations may be inspected for covert mobile missile activities.

Rail-Mobile ICBMs

In principle, rail-mobiles should be easier to verify than road-mobiles since they are restricted to rail lines, and the mobile ICBM trains should be more difficult to conceal. In practice, the Soviet Union has approximately 150,000 kilometers of heavily traveled track on which missile launch cars and launch support cars could be covertly deployed among other rail traffic. The key to detecting covert deployment would seem to be the U.S. ability to detect the launch car since the other launch support activities could easily be concealed in standard 24-meter rail cars that are used for most Soviet rail activity. The importance of being able to detect illegal rail launchers is underlined by the fact that ten covert launch cars would be equivalent to 100 covert road-mobile launchers because of the ten reentry vehicles on the SS–24 missile.

During the Moscow Summit, the sides discussed ways of reducing the rail-mobile verification problem, and some areas of agreement were reached on deployed rail-mobiles which parallel the cooperative measure for deployed road-mobiles. For example, deployed rail-mobile ICBMs and their launchers would be confined to a limited number of rail garrisons, which would be specified in the MOU. The number of shelters for rail-mobile trains would not exceed the number of trains specified for the garrison in the MOU. No shelter should be capable of holding more cars than the number on a standard missile train. This would require that the composition of a standard missile train be specified in the MOU. The number of rail entrances/exits at each garrison would also be limited.

The sides also agreed that missiles and launchers would be permitted to depart their garrisons for routine movements and exercise and operational dispersals, subject to notification requirements and limitations comparable to those for road-mobiles. Missile trains would be permitted to vary from their standard configuration during notified transit between declared locations, provided the variation is reported upon completion of the movement. There would be no restriction on the configuration of trains during dispersals.

The sides have not agreed on the size of rail garrisons or whether standard train configurations could be changed when conducting maintenance in the garrisons.

Apparently, the Moscow discussions of road- and rail-mobile verification produced the first meaningful movement on a very difficult verification problem. While conceptual frameworks have emerged, the challenge will be to develop cooperative measures that will enhance and supplement NTM monitoring of both types of mobile ICBMs.

Air-Launched Cruise Missiles (ALCMs)

The new Bear H heavy bomber is the only Soviet heavy bomber capable of launching their long-range, nuclear-armed AS–15 ALCM. This year, it will be joined by the Blackjack bomber, which is capable of carrying ALCMs and nuclear gravity bombs. Unless the projected force mix of missiles and bombers is changed because of a START agreement, by the mid-1990s, the U.S. projects that half of the Soviet heavy bomber weapons will be carried by the Blackjack and roughly 45 percent will be carried by the Bear H.[8]

The U.S. has converted 98 B–52Gs to carry up to 12 externally mounted ALCMs. B–52Hs are undergoing a similar conversion, which is scheduled for completion in fiscal year 1990. Eventually, the B–52H will be modified further to have a common strategic rotary launcher (CSRL) in the bomb bay for eight more ALCMs, eight nuclear-armed Short Range Attack Missiles (SRAMs), or a mix of both. The new B–1B can carry 12 ALCMs on underfuselage stores stations, with provisions for eight more carried internally on a CSRL. ALCM–equipped units are now located at seven Strategic Air Command (SAC) bases. In 1989 or 1990, the new air-launched advanced cruise missile (ACM), which is to be used on B–52Hs and B–1Bs, will begin initial deployment to an eighth base.[9]

Several factors complicate an already complex problem of verifying ALCMs on heavy bombers. First, the U.S. has nonnuclear (conventional bomb) B–52 squadrons and other non–ALCM B–52 units that retain the nuclear gravity bomb and SRAM mission. Second, under a START agreement, if the U.S. retains 4,900 ICBM and SLBM warheads, it would not be able to deploy more than 1,100 ALCMs. In order to optimize its nuclear delivery capability with only 1,100 ALCMs, the Air Force would likely opt for a mixed load of ALCMs, gravity bombs, and SRAMs, rather than loading each ALCM–carrying bomber to its maximum ALCM capacity — the latter would dictate an ALCM–carrier force of fewer than 100 heavy bombers. In practice, mixed rather than full ALCM loading is probably used now to maximize operational effectiveness. If so, an ALCM counting rule based on maximum capacity loading, which would be the most verifiable rule, would result in significant overcounting. Third, even with an ALCM counting rule of 10 ALCMs per heavy bomber, rather than 12 or 20 based on capacity loading, the Air Force will more than likely seek to reconvert some ALCM–capable B–52s to non–ALCM carriers rather than destroy the aircraft which would be required to reach agreed START limits. Some of those may be converted to nuclear gravity bomb and SRAM roles, and others may be converted to nonnuclear roles, which presumably would not be subject to treaty limits. Verification of such conversions will not be an easy problem to solve. Fourth, since NTM can only be used to monitor externally mounted ALCMs, there will be the continuing question of whether ALCMs are mounted on a rotary launcher in the bomb bay. Fifth, there will be the need to ensure that long-range, nuclear-armed ALCMs are not covertly carried by aircraft that are not declared to be heavy bombers. Finally, the U.S. is apparently interested in preserving the option to deploy long-range, conventionally armed ALCMs; thus, some means must be found to distinguish treaty-limited, nuclear-armed ALCMs from conventionally armed ALCMs, which would not be limited.

This overview should provide an appreciation of the scope of the problem that must be resolved in reaching agreement on ALCM counting rules for heavy bombers and in verifying compliance with the declared numbers of deployed ALCMs. It should be emphasized that the sides are seeking to limit only those ALCMs that are deployed on heavy bombers; neither side is seeking to limit nondeployed ALCMs (which would be virtually impossible to verify).

Prior to the Moscow Summit, the sides agreed on a counting rule for heavy bomber armaments according to which heavy bombers equipped only for nuclear gravity bombs and SRAMs will count as one delivery vehicle against the 1,600 limit on strategic nuclear delivery vehicles and one warhead against the 6,000-warhead limit. In addition, each long-range, nuclear-armed ALCM that is attributed to each type of heavy bomber by an agreed-upon counting rule shall count against the 6,000-warhead limit, and its heavy bomber shall count as one delivery vehicle.[10] They had also agreed to enhanced observation of activities by NTM, including open display of treaty-limited items at bomber bases and times chosen by the inspecting party.

In the Moscow Summit Joint Statement, it was announced that the sides achieved substantial common ground on ALCMs. During the summit, expert group discussions focused on counting rules for bombers, distinguishability rules for bombers and ALCMs, and some collateral measures.[11]

To assist in distinguishing among the different categories of heavy bombers, it was agreed that ALCM–equipped heavy bombers would be based separately from heavy bombers that are not ALCM–equipped, as well as heavy bombers equipped for nonnuclear roles. They also prohibited ALCMs at non–ALCM bases. Visits by the various categories of heavy bombers to bases of another category of heavy bomber must be notified to assist NTM monitoring.

The sides agreed the ALCM–equipped heavy bombers may be converted, through agreed procedures, to non–ALCM heavy bombers and to heavy bombers equipped for nonnuclear roles. They also agreed to conversion of a limited number of nuclear-armed heavy bombers to tanker, jamming, or reconnaissance aircraft, which would not be limited by the treaty. Furthermore, it was agreed that former heavy bombers that have been converted to tanker, jamming, or reconnaissance aircraft may be based at ALCM bases.

To reduce the distinguishability problems, it was agreed that ALCM–equipped heavy bombers will be distinguishable from other heavy bombers. The basis for this distinguishability is subject to further negotiation — and will require considerable ingenuity. They also agreed to count all currently existing long-range ALCMs as nuclear armed, and that future conventionally armed, long-range ALCMs would be distinguishable from nuclear-armed ALCMs. This implies that future ALCMs will not be dual capable, even though suspicions about warhead conversion will linger. In addition, the sides agreed that any long-range ALCM which has been tested and deployed with a nuclear variant may only be carried on ALCM–equipped heavy bombers.

While those areas of agreement are a step forward, the sides continue to disagree on the critically important ALCM counting rule. The U.S. proposes to attribute 10 warheads to each ALCM–equipped heavy bomber, while the Soviets propose that the number be based on the maximum number of ALCMs that a heavy bomber is equipped to carry. They also disagree on the range threshold that would be used to define "long range" and therefore be subject to treaty limits. The U.S. proposes 1,500 kilometers; the Soviets propose 600 kilometers. In addition, further negotiations will be required on what categories of bomber bases will be subject to on-site inspection, and whether conventionally armed heavy bombers would be counted against the 1,600 limit on delivery vehicles.

In sum, much has been agreed on and most of the remaining ALCM problems have been framed. On the other hand, the outlook for SLCMs is well described by the sole reference in the Moscow Summit Statement, "The sides also discussed the question of limiting long-range, nuclear-armed SLCMs."

Sea-Launched Cruise Missiles (SLCMs)

The new long-range, nuclear-armed SS–N–21 SLCM is now entering the Soviet inventory. The SS–N–21 is small enough to be fired from standard Soviet torpedo tubes and will probably be deployed on the Victor-class attack submarine. It could also be deployed on Akula-class and Sierra-class attack submarines. In addition, the Soviets are testing a larger, long-range, nuclear-armed SLCM, designated the SS–NX–24. This missile has been flight-tested from a specially converted Yankee-class nuclear-powered cruise missile attack submarine (SSGN).[12] Future variants of the SS–N–21 and SS–NX–24 could be accurate enough to permit the use of conventional warheads, depending on munitions development and the types of guidance systems incorporated in their design.

The most recent Soviet proposals of a limit of 400 long-range, nuclear-armed SLCMs on two classes of submarines and one class of surface ship and a limit of 600 long-range, conventionally armed SLCMs on one class of surface ship may be an indication of their deployment plans for those missiles.[13]

As with ALCMs, the Soviets have defined long-range SLCMs as missiles with a range of more than 600 kilometers. This range threshold would exclude six or seven types of older Soviet SLCMs. Most of those SLCMs are antiship missiles, and most are thought to be nuclear armed, presumably to compensate for poorer accuracy.

The United States has deployed three types of Tomahawk SLCMs that are observably identical: the long-range, nuclear-armed land-attack missile, designated TLAM/N; the long-range, conventionally armed land-attack missile, designated TLAM/C; and the shorter-range (450 kilometer), conventionally armed antiship missile, designated TASM. The Navy is planning to buy approximately 4,000 Tomahawk SLCMs through the 1990s. It is widely known that 758 of those missiles will be nuclear armed and that they will be deployed on approximately 200 surface ships and submarines.[14] As of March 1988, Tomahawk SLCMs were deployed on 32 Los Angeles–class and Sturgeon-class submarines and on 20 surface ships (battleships, nuclear cruisers, Ticonderoga class).

The Soviets have insisted that a START agreement must constrain the nuclear threat posed by long-range SLCMs. They have put forth a series of proposals beginning with their preference to ban all SLCMs, a ban on nuclear-armed SLCMs, a limit of 400 nuclear-armed SLCMs on two classes of submarines, and their most recent, previously described proposal.

The United States is unwilling to accept a SLCM ban because other navies have antiship SLCMs and because long-range SLCMs contribute to both nuclear and conventional deterrence. Even if nuclear-armed SLCMs are limited by START, the U.S. insists that conventionally armed SLCMs not be limited. Furthermore, agreement to the current Soviet SLCM proposal would require the U.S. to significantly alter its deployment plans regarding the numbers of nuclear-armed and conventionally armed SLCMs, and the types of ships and submarines that carry those SLCMs. The U.S. has consistently countered Soviet proposals for SLCM limits with the argument that the U.S. been unable to find an acceptable means of effectively verifying SLCM limits.

At the December 1987 Washington Summit, President Reagan and General Secretary Gorbachev agreed that they should find a mutually acceptable solution to limiting the deployment of long-range, nuclear-armed SLCMs. They agreed that the limitations on SLCMs would not be part of the 6,000-warhead limit or the 1,600-strategic nuclear delivery vehicle (SNDV) limit. Both sides committed themselves to establish ceilings on such missiles and seek mutually acceptable and effective methods of verification of such limitations, which could include the employment of NTM, cooperative measures, and on-site inspection.

Some form of limitation on long-range, nuclear-armed SLCMs under a START agreement would appear to be inevitable to satisfy both the Soviets and domestic critics who would ridicule a START agreement that permitted an unlimited number of nuclear-armed SLCMs. The dilemma is finding an acceptable and effective method of verifying the limitation.

Thus far, verification of cruise missile range, distinguishing nuclear-armed from conventionally armed SLCMs, verifying the number of SLCMs aboard surface ships and submarines, and verifying that nuclear-armed SLCMs are not covertly deployed on undeclared vessels have proven to be intractable problems. Whereas both sides have intelligence estimates of the other side's current and projected SLCM deployments, there are ample opportunities for undetected cheating under a START agreement that cannot be removed by verification provisions so far envisioned. That is not to say that the U.S. has given up on the problem, only that the outlook is not particularly bright.

The good news is that the sides have agreed to limit long-range, nuclear-armed SLCMs on ships and submarines rather than the total inventory (both deployed and nondeployed). A total inventory limit would be impossible to verify, since SLCMs can be concealed in virtually any building.

Scenarios to circumvent the range threshold, regardless of whether it is established at 600 kilometers or some other cutoff, are numerous. The concern, of course, is that a SLCM with a declared range below the threshold, and therefore not subject to agreed limits, could have a longer operational range. For example, the missile could be covertly flight-tested at an inland range to minimize the risk of detection. There is also no overriding reason that the missile be flown to its maximum range, only that it demonstrate that it can perform effectively in all flight phases (e.g., launch, climb, cruise, terminal approach, and impact). Performance reliability to fuel depletion could be confirmed on a test stand. The range of a shorter-range SLCM could also be extended by replacing its heavier conventional warhead with a nuclear warhead, minimum ballast, and more fuel. More sophisticated cheating scenarios are possible; however, the point should be clear that suspected violations of range capability will be exceedingly difficult to prove. This will place a premium on the U.S. intelligence collection and analysis capabilities.

The problems involved in distinguishing nuclear-armed missiles from conventionally armed missiles have received considerable media attention during the last year. More accurately, the challenge is to find a means of determining that a conventionally armed SLCM, which the U.S. would not limit, is not a nuclear-armed SLCM, which would be limited. There is also serious concern about the replacement of the conventional warhead with a nuclear warhead during crisis. The replacement problem goes well beyond what verification is expected to do, since the SLCM would be a legal, conventionally armed SLCM until its warhead was replaced with a nuclear warhead — and START limits would then become academic.

The well-publicized Soviet proposal for distinguishing between nuclear and conventionally armed SLCMs has been analyzed and rejected by the U.S. as ineffective. Lt. Gen. Colin Powell, White House National Security Advisor, in a speech before the Atlantic Council in June 1988, said "The Soviet technology for verifying nuclear SLCMs requires that inspectors be extremely close to the source of the radiation. It is also easy to cheat. There is no technical means that can't easily be defeated."

U.S. scientists have suggested a number of possible solutions to the distinguishability problem, such as using neutron scanners which would emit gamma rays from a nuclear warhead that could be detected; placing tamper-proof seals on conventionally armed SLCMs where the warhead section is mounted on the missile; attaching radiation-sensitive dosimetric tags on conventionally armed missiles; and sealing the warhead to the missile with an electronic lock.[16] All have been rejected because they could be circumvented, were overly intrusive, would interfere with routine maintenance requirements, and so forth. Presumably, scientists on both sides are continuing to search for an acceptable method of counting nuclear-armed SLCMs.

Several factors add to the challenge facing the scientific and policy communities in seeking an acceptable method of verifying the number of nuclear-armed SLCMs aboard a vessel. First, there will be the requirement to distinguish U.S. nuclear-armed SLCMs from conventionally armed land-attack and antiship SLCMs, which are observably identical. Second, most U.S. vessels that have nuclear-armed SLCMs also have other types of nuclear-armed weapon system (e.g., Terrier air defense missiles, ASROC or SUBROC antisubmarine weapons and depth bombs).[17] Thus, unless close access to each weapon system is permitted, which is unlikely, nuclear warhead scanners could only determine the number of warheads and not the number of nuclear SLCMs. Heavily shielded storage areas could also be used on larger surface ships to protect concealed nuclear SLCMs from nuclear scanners. Third, attributing some number of nuclear-armed SLCMs to each armored box launcher or vertical launch system aboard ships is complicated by the fact that they are capable of launching other conventionally armed and nuclear-armed weapons. Furthermore, counting launchers would be ineffective, since the potential number of SLCMs aboard each ship would only be bounded by its storage capacity. Fourth, even if the number of nuclear-armed SLCMs aboard each SLCM–capable vessel was confirmed by on-site inspection prior to departure from port, concerns would remain about covert loading beyond the port. Boarding ships at sea would likely be objectionable to the U.S. and probably totally unacceptable for attack submarines for operational reasons.

National Technical Means would be the primary method of detecting covert nuclear-armed SLCMs on undeclared vessels. The key indicator on surface ships would, of course, be the launcher. In addition to looking for concealed launchers, analysts could be expected to closely monitor the configuration of non–SLCM launchers to ensure that they have not been modified for long-range SLCMs. If the Soviets did not believe that testing was necessary to confirm the SS–N–21 launch characteristic from an undeclared class of attack submarine, the chances of NTM detecting covert SLCM deployments would be greatly reduced. Because use of attack submarines for a long-range SLCM mission would be at the expense of their primary mission of defending SSBNs, analysts would likely be alert for increased attack submarine production rates.

Compliance judgments relating SLCM deployments on undeclared vessels will probably be based on NTM alone, since it is very unlikely that on-site inspection would reveal a "smoking gun." Inspections at sea would make transferring illegal SLCMs less attractive; however, it is unlikely the U.S. Navy would accept boardings at sea. Even if such boardings were acceptable, as a last resort, the illegal missiles could be dumped overboard.

This less-than-encouraging outlook is complicated further by a completely different policy issue that must also be resolved in formulating some acceptable way of limiting and verifying nuclear-armed SLCMs. The longstanding U.S. policy of neither confirming nor denying the presence of nuclear weapons on naval vessels will obviously preclude declaring in the MOU that nuclear-armed SLCMs are carried aboard specific vessels. There are no indications that the neither-confirm-nor-deny policy will be abandoned because of the SLCM problem. To the contrary, the suspension of military obligations to New Zealand in 1986, the current Danish resolution banning port calls by nuclear armed vessels, and a similar bill in the Philippine House and Senate emphasize the importance of retaining the policy.

In sum, potential circumvention options undermine U.S. ability to effectively verify limits on long-range, nuclear-armed SLCMs with any type of verification regime that has been envisioned. Unless a breakthrough occurs, some variation of "each side merely declaring its planned number of nuclear-armed SLCMs," which has been suggested, may be as reasonable as any other approach. Compliance judgments could be based on the same intelligence estimates that are used for threat assessment in the absence of arms control limits. Though not an ideal outcome, such an approach would acknowledge the need to constrain nuclear-armed SLCMs and the fact that the constraint cannot be effectively verified. Safeguards against breakout could be developed to reduce the risks inherent in this type of declaratory approach.

Conclusions

The U.S. proposal to ban mobile ICBMs because of verification difficulties appears to be on the slippery slope, particularly if agreement can be reached on verification regimes that would reduce the uncertainties associated with verifying limits on deployed and nondeployed road- and rail-mobiles. The verification regime that the sides have agreed to use for nondeployed silo-launched ICBMs and SLBMs could also be used for nondeployed mobile ICBMs and their launchers.

Because the sides must rely heavily on NTM to monitor limits on deployed road- and rail-mobiles, they have agreed to bound the monitoring problem by confining the mobiles to small restricted areas and rail garrisons, with several exceptions for notified movements beyond those areas. Nevertheless, improved NTM capabilities and coverage will be necessary, and will likely be funded, to enhance the U.S. ability to detect covert mobiles outside declared locations or areas. Extensive cooperative measures to enhance NTM and on-site inspection to supplement NTM will also be needed. Whereas verification uncertainties will persist for road- and rail-mobiles, those uncertainties must be reduced to acceptable levels of risk.

Two major problems must be resolved in limiting and verifying ALCMs: the number of ALCMs to be attributed to heavy bombers (i.e., a counting rule) and the requirement to distinguish ALCM–carrying heavy bombers from other heavy bombers.

The ALCM counting rule involves a tradeoff between verification and operational loadings. The most verifiable counting rule would be based on maximum loading capacity; however, this would result in significant overcounting. To preserve operational flexibility, a counting rule will be needed that more closely approximates actual ALCM loadings.

The sides have apparently agreed to adopt a separate basing scheme to resolve the distinguishability problem for ALCM carriers. This would result in separate bases for heavy bombers with ALCMs, heavy bombers with nuclear gravity bombs and SRAMs, and heavy bombers with conventionally armed weapons. A series of cooperative measures that support this scheme have also been agreed upon, with only a few outstanding issues.

Thus far, the sides have been unable to find an acceptable and effective method of verifying limits on long-range, nuclear-armed SLCMs. Potential circumvention options undermine U.S. ability to effectively verify the maximum range of SLCMs, to confirm that a conventionally armed SLCM is not nuclear-armed, and to verify the number of nuclear-armed SLCMs aboard surface ships and submarines.

Even though the sides continue to seek an acceptable solution to the SLCM problem, the outlook is not particularly promising. Unless a breakthrough occurs, a declaratory approach in which each side merely declares its planned number of nuclear-armed SLCMs aboard ships and submarines may be as reasonable as any other approach. In practice, this approach would be essentially "risk free" for the Soviets, since they have full access to U.S. procurement plans for nuclear-armed SLCMs before the Congress. On the other hand, the U.S. would have to base its compliance judgments on intelligence estimates rather than data from a verification regime. Appropriate U.S. safeguards could serve as a hedge against the risks of breakout under such an approach.

In sum, mobile ICBMs, ALCMs, and SLCMs are serious obstacles to a START agreement. The U.S. is unwilling to ban ALCMs or SLCMs, and the Soviets are unwilling to ban mobile ICBMs. Thus, some other means of constraining those systems will be necessary. While both sides continue to seek ways to reduce the monitoring uncertainties involved in verifying numerical limits on mobile ICBMs, ALCMs, and SLCMs, it is very unlikely that a preferred level of monitoring confidence will be achieved. Accordingly, if a START agreement is to be concluded and ratified, the U.S. must be prepared to accept the monitoring uncertainties that accompany limits on mobile ICBMs, ALCMs, and SLCMs.

Notes

1. U.S. Department of Defense, *Soviet Military Power* (Washington, DC: U.S. Government Printing Office, April 1988), pp. 15, 47.

2. *Ibid.*, p. 46.

3. U.S. Congress, House, *MX Rail Garrison and Small ICBM: A Program Review*, Report of the Subcommittee on Research and Development and Subcommittee on Procurement and Military Nuclear Systems of the Committee on Armed Services, March 1988, pp. 7-14.

4. Dante B. Fascell, *Congressional Record*, March 3, 1988, pp. E522-E525.

5. See *Joint Statement Between the United States and the Union of Soviet Socialist Republics* issued following meetings in Moscow, May 29–June 1, 1988, pp. 4–5.

6. U.S. Congress, House, *Breakout, Verification, and Force Structure: Dealing With the Full Implications of START*, Report of the Defense Policy Panel of the Committee on Armed Services, May 1988, p. 40.

7. Most areas of agreement and disagreement on mobile ICBMs are reported in John S. McCain, *Congressional Record*, June 23, 1988, pp. S8508–S8514.

8. *Soviet Military Power, op. cit.*, pp. 50–53.

9. "Strategic and Tactical Nuclear Missiles," *Air Force Magazine*, May 1988, p. 188.

10. Fascell, *op. cit.*, p. E523.

11. Most areas of agreement and disagreement on ALCMs are reported in McCain, *op. cit.*, pp. S8510–S8511.

12. U.S. Department of Defense, *Soviet Military Power* (Washington, DC: U.S. Government Printing Office, March 1987), pp. 37–38.

13. Fascell, *op. cit.*, p. E524.

14. Nuclear Notebook, *Bulletin of Atomic Scientists*, June 1988, p. 56.

15. Nuclear Notebook, *Bulletin of Atomic Scientists*, July/August 1988, p. 55.

16. Nancy Cooper and Douglas Waller, "Inspect Unto Others," *Newsweek*, March 28, 1988, p. 42.

17. Nuclear Notebook, June 1988, *op. cit.*, p. 56.

21

Low-Threshold Test Ban
Is Feasible*

*Harold A. Feiveson, Christopher E. Paine,
and Frank von Hippel*

I n February 1987, the Reagan Administration restated its position on nuclear testing as follows: "As long as we depend on nuclear weapons for our security, we must insure that those weapons are safe, secure, reliable and effective. This demands some level of underground nuclear testing as permitted by existing treaties." This policy statement does not, however, indicate the frequency and yields of test that these objectives would require.[1]

It is our contention that acceptable standards of weapon safety, security, and reliability for the nuclear arsenal could be maintained under a low-threshold test ban treaty (LTTBT) that prohibited all tests except those below 1 kiloton plus a small number of tests in the 5-to 15-kiloton range. This position is shared by a number of former high-level weapons designers.[2]

In this article, we discuss the verifiability of a 1-kiloton threshold test ban with a quota of above-threshold tests and the impact of such a ban on tests for weapons safety and security, reliability, and weapons effects. We then discuss the opposing positions on the development of more "militarily effective" nuclear weapons — the principal real issue dividing test-ban advocates and opponents.

*Reprinted from Science, Vol. 238, pp. 455-459, 23 October 1987. Copyright 1987 by the AAAS.

Verification

Under a LTTBT, each country would be permitted to test only within the confines of a single designated area. The detection of a nuclear explosion of any magnitude elsewhere would therefore be *prima facie* evidence of a violation.

There is now general agreement within the expert community that existing external networks of high-performance teleseismic stations have the capability to detect and identify ordinary underground nuclear explosions in hard rock down to 1 kiloton anywhere in the Soviet Union and well below that in some areas.[3] However, it has been known since 1959 that it is possible to muffle or "decouple" small nuclear explosions in large underground caverns. In a cavity with a radius large enough so that the energy of the shock wave could be absorbed by elastic deformation of the rock, the apparent yield of an underground nuclear explosion could be reduced by a factor of about 100.

Such "full decoupling" of even low-yield explosions would be a difficult and uncertain task. Leakage of radioisotopes would have to be prevented, and the cavity would have to be protected against collapse subsequent to the blast to avoid creating a telltale subsidence crater at the surface. The full decoupling of a 5-kiloton explosion would require a cavity 60 to 90 meters in diameter (big enough to contain a 20- to 30-story building).[4] Because the volume of the cavern required increases in direct proportion to the yield of the nuclear explosion, it is generally agreed that full decoupling would be completely impractical for yields above 10 kiloton.

Nevertheless, since decoupling in the 1- to 10-kiloton yield range cannot be entirely ruled out as an evasion technique, verification of any treaty banning nuclear explosions with less than 10-kiloton yield would require internal, as well as external, seismic monitoring stations. It appears to be generally agreed that, with 25 to 30 carefully sited seismic stations within the Soviet Union, even fully decoupled nuclear explosions could be reliably detected and identified down to yields of a few kilotons. Using the fact that, like small explosions, decoupled ones radiate a much larger fraction of their seismic energy at high frequencies, some seismologists argue that a network equipped with high-frequency (5 to 50 Hz) seismometers could detect and reliably distinguish from earthquakes decoupled nuclear explosions down to approximately 1 kiloton.[4]

In regions containing rock suitable for large decoupling caverns, however, arrangements would be needed to verify that large chemical explosions involving tens of tons of explosives were not decoupled, low-kiloton nuclear explosions. This would probably require prior notification of major industrial and mining explosions and occasional inspections of the sites of such events.

There would be little pressure to try to carry out clandestine decoupled explosions under the low-threshold treaty being discussed here. The most important benefits that could be derived from these explosions would be legally available through the quota of test explosions of up to 15 kiloton.

Verification of a low-threshold test ban will also require the capability to estimate the yields of nuclear explosions at the designated test sites to assure that they do not exceed the agreed threshold. For improved verification of the present 150-kiloton threshold test ban, the Reagan Administration has advocated use of the CORRTEX (Continuous Reflectometry for Radius versus Time Experiment) method, which would measure the speed of the strong shock wave near the explosion by means of a cable placed in a satellite hole 10 to 15 meters from the weapon emplacement hole.[5] This proposal has caused some concern in the U.S. weapons laboratories because it would be relatively intrusive and require restrictions on the size and contents of the canisters containing the nuclear device and diagnostic equipment.[6] For a 1-kiloton threshold, such restrictions would become so stringent and the CORRTEX cable would have to be brought so close to the explosive (2 to 3 meters) that the technicians from the two sides would have to work virtually as one team. Seismic yield estimation techniques will therefore be required.

Seismic yield verification for 1- to 10-kiloton explosions would require in-country seismic stations. The accuracy of the measurements would be improved by requiring that permitted tests be carried out only in strong-coupling (water-saturated, for example) media located within the one small designated test area, and by calibrating the seismometers with nuclear explosions of independently determined yield. A modest degree of on-site inspection would be required to verify that the designated test area did not afford opportunities for significant decoupling. Even lacking assurance of compliance with such arrangements, the uncertainty of yield could probably be kept to within a factor of two at one kiloton and 50 percent at the quota threshold of 15 kiloton (95 percent confidence level). Given adequate assurance of compliance with these arrangements, any extended series of tests exceeding the threshold by 20 to 30 percent would be detected.[7] The quota of 5- to 15-kiloton tests would greatly reduce any incentives to cheat at the margins of the 1-kiloton threshold.

The weapons labs and the Department of Energy have recently argued that the Soviet Union might conduct clandestine tests in deep space — behind the sun, for example. We relegate our comments on this scenario to a footnote.[8]

Safety and Security

Another technical reason often given for continued testing is the need to improve the safety of nuclear weapons and their security against unauthorized use.

After four decades of development, the safety design of nuclear weapons is well advanced. In particular, all U.S. nuclear weapons are said to be "one-point safe;" that is, they are designed not to produce a significant nuclear yield even if the chemical explosives are triggered at one point by the penetration of a bullet or by fire. And many U.S. weapons have environmental-sensing devices, for example, which sense acceleration and altitude and block triggering signals from reaching the chemical explosives, unless the weapon has gone through its intended launch-to-target trajectory. Such systems do not require nuclear testing.

Recent work on safety improvements has been focused on the much less serious problem of reducing the probability of dispersal of toxic plutonium in an accident. An important advance in this regard has been the use since 1980 of "insensitive high explosives" (IHE), which are less susceptible to detonation in abnormal situations such as fires or aircraft crashes. Warheads containing IHE are now available for high-yield and low-yield bombs, all U.S. cruise missiles, the Pershing II, and the MX.[9] In cases where new warheads containing IHE have not been developed, there is usually no intention to do so for institutional or technical reasons.[10-12]

Most other improvements do not require certification by a nuclear explosive test. For example, mechanical or electrical design improvements that do not alter the geometry of the fissile material or chemical implosion mechanism in the weapon are ordinarily tested by removing the fissile material and replacing it by non-chain-reacting material such as uranium–238. More sensitive tests of the degree of compression that has been achieved by the chemical implosion are conducted by removing only a part of the fissile material, leaving enough to produce a yield equivalent to a very small nuclear explosion of less than $10(-6)$ kiloton. Measurements of the production of neutrons from such "zero-yield" nuclear tests were, in fact, used by the United States to explore safety problems during the 1958-1961 U.S.–Soviet nuclear testing moratorium.[13]

Permissive action links (PALs), the electronically coded locks that are used to secure U.S. nuclear weapons from unauthorized use, have already gone through several generations of improvements. The primary issue today is not further technical refinement but rather the fact that many weapons in the U.S. stockpile, including the weapons on ballistic-missile submarines, still have no PALs at all.[14] A test ban would not prevent the introduction of modern (category D, six-digit code) PALs into currently unprotected weapons or weapons with earlier generation PALs because this type of PAL works on components that do not require nuclear tests to certify their performance.[14]

Reliability

Concerns raised by the weapons labs that confidence in the reliability of the U.S. nuclear stockpile could not be established in the absence of testing played a key role in derailing President Carter's efforts to achieve a Comprehensive Test Ban.[15] The technical basis for this concern was immediately challenged in a letter to Carter from a former weapons laboratory director and two more former weapon designers[16] and the debate has continued among the experts ever since.[17]

It is difficult for outside observers to reach a conclusion on the technical aspects of the stockpile confidence issue based on such fragments of the debate as have been declassified. However, the public record does support three important findings:

First, as the Department of Energy has acknowledged, weapon designs which are reliable enough to be manufactured without statistically significant numbers of nuclear explosive proof-tests are also reliable enough to be remanufactured in the future.[18] The issue of warhead reliability therefore concerns the rare case of the appearance of novel design or material flaws that cannot reasonably be remedied by restoring the weapon to original specifications.

Second, because nuclear tests are expensive, only a small number of nuclear explosive tests of stockpiled weapons have been conducted to resolve reliability issues. The principal way in which problems in the stockpile are detected and rectified is by disassembly and inspection, and by nonnuclear tests. During the period 1970–1985 only six to eight underground nuclear explosions were justified by the need to "correct defects in stockpiled weapons."[19] A comparable number may have been carried out to determine the seriousness of problems detected during routine disassembly and inspection. The resulting average of about one "stockpile-confidence" test per year should be compared with the average of 16 U.S. nuclear tests per year during this same period.[20]

Third, to the limited extent that reliability problems have arisen in thermonuclear weapons, apparently virtually all have occurred in their miniaturized fission triggers.[21]

Judging from the high relative frequency of U.S. tests in the yield range 5 to 15 kiloton (nearly 40 percent of all U.S. tests during 1980–1984)[22] and our own calculations,[23] the yields of the triggers for U.S. thermonuclear weapons appear to fall in the range 5 to 15 kiloton. If future changes in stockpiled thermonuclear weapons were confined to conservative modifications of existing trigger designs, a quota of about one test per year at a yield of about 5 to 15 kiloton could therefore satisfy the concerns that have been raised about the need for reliability tests. An independent review with full access to the relevant information might well establish that even this small number of tests could be phased out within a few years, if no significant changes were introduced into the weapons stockpile.

Nuclear Weapons Effects

One rationale for continuing underground nuclear explosions that has received increasing public emphasis in recent years has been the need to examine the ability of military equipment — including nuclear warheads and their reentry vehicles — to withstand the effects of nearby nuclear explosions. However, since most of the knowledge obtainable from underground tests can be obtained with explosions with yields of less than 1 kiloton, the need for "weapons-effects" tests is not a strong argument against an LTTBT. Indeed, for this reason and because tests involving smaller yield explosions are less expensive, most U.S. nuclear weapons effects tests are already conducted at quite low yields.[24] The permitted quota of higher-yield tests could be used for those few applications where a higher-energy spectrum of x-rays would be advantageous.

Do We Need New Types of Nuclear Weapons?

A major benefit to the U.S. of more stringent limits on the testing of nuclear weapons would be to impede the development of new nuclear weapons by the

Soviet Union. This benefit is, however, scarcely mentioned by government and laboratory officials involved in the test ban debate. Instead, in their congressional testimony, they reiterate their concern that additional testing restrictions would impede their own work on the development of new nuclear weapons.

For example, in 1985, C. Paul Robinson, then principal associate director for National Security Programs at Los Alamos National Laboratory, argued as follows[25]: " ... [a test ban] would prevent us from validating the development of weapons that would allow us to respond to new requirements such as those which may derive from the changes that are occurring in the targets we must hold at risk in the Soviet Union. These requirements might include... developing earth penetrating weapons to hold at risk extremely hard, buried targets (missile silos, deep underground facilities) and developing effective means to hold at risk mobile and imprecisely located targets ...." Robinson suggested that one way to incapacitate Soviet mobile weapons systems might be with very high levels of microwave radiation generated by a specially designed nuclear weapon. This "third-generation" nuclear weapon concept is now being actively researched at the weapons laboratories — as is the nuclear explosion-pumped x-ray laser for attacking satellites and ballistic missiles in space. Indeed, nuclear directed-energy weapons consume about one half of the U.S. budget for exploratory research on nuclear weapons, and the remaining half is primarily focused on improving U.S. capabilities to attack Soviet nuclear forces.[26]

Another frequently claimed benefit of continued testing — reduction in the destructiveness of nuclear arsenals — was recently cited in a White House strategy document as follows[27]: " ... the United States does not target population as an objective in itself and seeks to minimize collateral damage through more accurate, lower yield weapons." In fact, despite dramatic increases in accuracy, the W–87 warhead for the MX missile has twice the yield of the original warhead on the Minuteman III missile which it is replacing, and the yield of the W–88 warhead for the Trident II missile is about ten times as great as the yield of the warhead on the submarine-launched Poseidon ballistic missile.[28]

Despite weapons "modernization" to increase the "credibility" of nuclear war fighting postures, the foundation of stable deterrence will continue to be provided by the inescapable mutual vulnerability of the United States and Soviet Union to nuclear attack. Weapons modernization is not only wasteful of resources and scientific talent, however. It is also dangerous. Its justification within both countries demands exaggerated and worst-case caricatures of the adversary's intentions and capabilities and continually reinforces dehumanizing images of the opposing national leadership. Also, the nuclear war-fighting systems that have resulted — for example, the Soviet heavy SS–18 ICBM and the U.S. MX and Trident II — could increase fears of preemptive strikes, undermining restraint in times of crisis.

Effects of a Low-Threshold Test Ban

A 1-kiloton threshold test ban would severely impede the development of all new nuclear missile warheads, bombs, and nuclear directed-energy weapons other than those with yields of a few kilotons or less.[29] To the extent that a small quota of tests with yields of up to 15 kiloton were exploited for weapons development rather than reliability and weapons-effects tests, some slow progress might also be made on the development of qualitatively new types of nuclear weapons with yields of tens of kilotons. This is to be contrasted, however, with the current situation of unlimited testing at a yield up to 150 kiloton — making possible the development of new types of nuclear weapons with yields up to about 500 kiloton.

However, a low-threshold test ban would not by itself prevent development and deployment of new strategic and tactical nuclear delivery systems. New delivery systems could be developed with nuclear warheads and bombs as a fixed rather than variable, parameter in their design. For example, the already-tested M–X warhead could be mated to the mobile, single-warhead Midgetman missile.[30] Obviously, if optimizing the Midgetman's capability to destroy Soviet strategic nuclear forces and their command facilities is the goal, such a solution may be "suboptimal" relative to what could be achieved with a new warhead. But, if the purpose of the Midgetman is to improve the survivability of the U.S. strategic nuclear forces, an LTTBT would not be a serious impediment. Indeed, it would forestall developments by the Soviet Union, similar to those underway in the U.S. weapons laboratories, of 26 nuclear warheads designed specifically for attacks on "strategic relocatable targets" (the Midgetman, for example).

Conclusion

A low-threshold test ban would be an important first step toward redirecting the vast bureaucratic and technical establishments that have been built on the illusion that nuclear weapons can be targeted and employed like other kinds of weapons to achieve traditional military goals. This misguided belief, in turn, sustains the illusion that endless weapons modernization is the key to national security.

At the same time, an LTTBT with a small quota of 5- to 15-kiloton tests would meet many, if not all, of the technical concerns raised by the weapons laboratories regarding the reliability, safety, and security of the stockpile and the need to harden critical military systems against weapons effects. It would also allow the weapons laboratories to maintain sufficient expertise to be able to respond to unexpected developments, including the breakdown of the treaty.

The low-threshold test ban would not, unfortunately, provide a guarantee against the development of possibly exotic new types of low-yield weapons and the exploration of the underlying physics and technology that could be used to develop higher-yield weapons if the treaty limits were to break down. It would also not have as much direct impact on nonproliferation as would a more comprehensive ban. For these reasons, some arms control experts advocate still more stringent limits. Richard Garwin, for example, would prefer a treaty that would allow "explosive releases of nuclear energy taking place only in permanently occupied above-ground buildings ...."[31] This might be taken to be a reasonable definition of a comprehensive test ban.

We share the hope that, after in-country seismic monitoring systems are fully established and tested, it will become politically possible to lower the threshold to below 1 kiloton and eventually to near zero as a result of increased public confidence in nonseismic means of verification. Only then would the nuclear-testing nations be in a position to present the treaty for signature by nonweapons states — thereby obtaining a meaningful technical barrier to the proliferation of thermonuclear weapons and an additional moral and political barrier to the spread of all nuclear weapons.

Notes

1. *New York Times*, 27 February 1987, p. A3.

2. Letter to Senators Kennedy, Hatfield, and DeConcini signed by H.A. Bethe, N.E. Bradbury, R.L. Garwin, J.C. Mark, G.T. Seaborg, and T.B. Taylor. Harold Brown, a former director of the Livermore National Laboratory and Secretary of Defense during the Carter Administration, although unwilling to align himself with any particular test ban proposal, has stated: "I can support an agreement to limit nuclear tests to a few a year at 10 to 15 kiloton and all others to 1 to 2 kiloton" (private communication to C.E. Paine, 5 May 1987).

3. See for example, W.J. Hannon, *Science*, 277, 251 (1985).

4. J.F. Evernden, C.B. Archambeau, E. Cranswick, *Rev. Geophys.* 24, 143 (1986); see p. 149.

5. "Verifying Nuclear Testing Limitations: Possible U.S.–Soviet Cooperation," Special Report No. 152 (U.S. Department of State, Washington, DC, 1986).

6. See for example, R.E. Batzel, Lawrence Livermore National Laboratory, prepared statement, hearing before the Senate Foreign Relations Committee, 15 January 1987, p. 16.

7. C.B. Archambeau, *Proceedings of the SIPRI/CIPPS Symposium on the Comprehensive Test Ban: Problems and Prospects*, Ottawa, Canada, 23 to 25 October 1986 (Oxford Univ. Press, New York, in press), pp. 4 and 14-15. See also, Archambeau, letter to Senators Kennedy, Hatfield, and DeConcini, 22 April 1987, and private communication.

8. See, for example, the testimony of Livermore Laboratory Director R.E. Batzel, in *Review of Arms Control and Disarmament Activities*, House of Representatives Committee on Armed Services (U.S. Government Printing Office, Washington, DC, 1986), p. 125. Tests in space would have to be done at great range and expense to evade detection by near-earth satellite sensors. If such cheating is truly of concern, the United States could deploy deep-space sensors — or work toward an agreement on nonintrusive prelaunch inspection of space payloads for the presence of nuclear weapons. The Outer Space Treaty forbids the placing of nuclear weapons in space and the Soviet Union has recently expressed a willingness to join in an agreement involving prelaunch inspection for weapons of all types (Ambassador Y. Nazarkine at the Conference on Disarmament, Geneva, 17 March 1987).

9. T.B. Cochran, W.M. Arkin, M.M. Hoenig, *U.S. Nuclear Forces and Capabilities* (Ballinger, Cambridge, MA, 1984), pp. 65, 200, 79, 182, 297, 126, and 133.

10. For example, the Navy has elected not to put IHE in the warhead for its Trident II ballistic missile because it believes that the safety advantages of IHE are not worth the weight penalty associated with its use. If the Navy changed its mind, the Trident II could be adapted to use the same warhead as the M–X. In the case of artillery shells, their small diameter makes them difficult to convert, since a larger volume of IHE is required to release a given amount of energy. Finally, replacement warheads are not being developed for some tactical weapons that are being phased out in favor of precision- guided conventional weapons.

11. See S. Fetter, in *The Comprehensive Nuclear Test Ban: For and Against* (Ballinger, Cambridge, MA, in press).

12. According to the official in charge of DOE's defense programs, "It is DOE policy to incorporate IHE in all warheads under development unless it is determined that use of IHE would cause unacceptable operational penalties for the Department of Defense deliver system, or that its use would simply not be possible for some technical reason....Warheads which do not incorporate IHE are designed to prevent nuclear yield in both normal and abnormal environments and are not considered unsafe. The DOE has not provided, and will never provide, to DOD a warhead which is considered unsafe." See S. Foley, "Responses to questions for DOE budget hearing," Subcommittee on Procurement and Military Nuclear Systems, House Armed Services Committee, February 24, 1987 (U.S. Government Printing Office, Washington, DC, in press).

13. See R.N. Thorn and D.R. Westervelt, *Hydronuclear Experiments* (Report LA–10902–MS, Los Alamos National Laboratory, Los Alamos, NM, 1987).

14. See, for example, T. Julian, in *Preventing Nuclear Terrorism*, P. Leventhal and Y. Alexander, Eds. (Lexington Books, Lexington, MA, 1987), pp. 180-181.

15. J. Carter, *Keeping Faith* (Bantam Books, New York, 1982), p. 229.

16. N.E. Bradbury, R.L. Garwin, J.C. Mark, letter to President Carter, reprinted in *Effects of a Comprehensive Test Ban Treaty on United States National Security Interests*, Hearings before the Panel on the Strategic Arms Limitation Talks and the Comprehensive Test Ban, House Committee on Armed Services (U.S. Government Printing Office, Washington, DC, 1978), p. 181.

17. See, for example, the report by J.W. Rosengren (*Some Little-Publicized Difficulties with a Nuclear Freeze* [Report RDA–TR–122116–001, R&D Associates, Marina Del Ray, CA, 1983]) and the critique of this report by R.E. Kidder (*Evaluation of the 1983 Rosengren Report from the Standpoint of a Comprehensive Test Ban* [Report UCID–20804, Lawrence Livermore National Laboratory, Livermore, CA, 1986]) and subsequent exchanges (J.W. Rosengren, Stockpile Reliability and Nuclear Test Bans: A Reply to a Critic's Comments [Report RDA–TR– 35822–001, R&D Associates, Arlington, VA, 1986] and R.E. Kidder, Stockpile Reliability and Nuclear Test Bans: Response to J.W. Rosengren's Defense of His 1983 Report [Report UCID–20990, Lawrence Livermore National Laboratory, Livermore, CA, 1987]).

18. According to DOE Assistant Secretary for Defense Programs S. Foley, "New warhead or bomb military characteristics submitted by the Department of Defense for acceptance by the Department of Energy normally contain a requirement that the design, development, and production of the warhead (or bomb) be well documented and involve processes that to the extent possible allow replication of the warhead (or bomb) at a future date. Assuming, therefore, that vendor-supplied materials and com-

ponents are still available at the time desired for remanufacture (and this will not necessarily be the case), the remanufacture of existing, well-tested warheads is possible." Department of Energy, National Security Programs Authorization Act for Fiscal Years 1987 and 1988, hearings before the House Committee on Armed Services (U.S. Government Printing Office, Washington, DC, 1986), pp. 127-128.

19. Eight tests were cited as necessary in Department of Defense-Arms Control and Disarmament Agency-Department of Energy, joint answer to a question for the record in Nuclear Testing Issues, hearing before the Senate Armed Services Committee (U.S. Government Printing Office, Washington, DC, 1986), p. 46. However, on 17 April 1986, in answers provided in writing to questions from Representative E. Markey, S. Foley stated that only six tests were required for this purpose.

20. *World Armaments and Disarmaments*, SIPRI Yearbook, 1986 (Oxford Univ. Press, New York, 1986), p. 129.

21. D.R. Westervelt, in *Proceedings of the SIPRI/CIPPS Symposium on the Comprehensive Test Ban: Problems and Prospects*, Ottawa, Canada, 23 to 25 October 1986 (Oxford Univ. Press, New York, in press).

22. R.E. Kidder, in *Proceedings of the Department of Energy Sponsored Cavity Decoupling Workshop*, Pajaro Dunes, California, 29-31 July 1985 (Report CONF–850779, Department of Energy, Washington, DC), p. V25.

23. F.N. von Hippel, H.A. Feiveson, C.E. Paine, *International Security*, in press.

24. R.S. Norris, T.B. Cochran, W.M. Arkin, *Known U.S. Nuclear Tests, July 1945 to 16 October 1986* [Report 86-2 (rev. 1), Natural Resources Defense Council, Washington, DC, 1986].

25. C.P. Robinson, in *Review of Arms Control and Disarmament Activities*, hearings of the House Armed Services Committee (U.S. Government Printing Office, Washington, DC, 1985), pp. 140-142.

26. The proposed $2.1-billion budget for U.S. nuclear weapons research and development for fiscal year 1988 includes $480 million for nuclear directed-energy weapons (about one-quarter of the total budget), $428 million for other future weapons technology development and testing (including work on "strategic relocatable targets" and "hard target kill"), and $424 million on warhead development engineering, testing, and certification. The remainder of the budget is for Nevada test site preparation, plant and equipment, and inertial fusion research. See also Congressional Budget Request: Atomic Energy Defense Activities, vol. 1, FY 1988 (U.S. Department of Energy, Washington, DC, 1987) and Energy and Water Development Appropriations for FY 1988, hearings before the House Committee on Appropriations (U.S. Government Printing Office, Washington, DC, 1987), part 6, pp. 643-736.

27. "National Security Strategy of the United States," The White House, January 1987, p. 21.

28. See pp. 116, 121, 137, and 145, in 9.

29. J.C. Mark in *Public Interest Report* (Federation of American Scientists, Washington, DC, December 1986), p. 12.

30. Former weapons designers R.L. Garwin, J.C. Mark, and H.A. Bethe have wryly observed that, "It might need shock- alleviation mounting for a mobile Midgetman subject to nuclear attack, but the demand for a new warhead is analogous to requiring

that one redesign an astronaut before launching him or her into space. Careful attention to packaging will do" (ibid. [April 1987], p. 11).

31. R. Garwin, in ibid. (December 1986), p. 13.

22

Facing Nuclear Reality:
The Test-Ban Issue*

George H. Miller, Paul S. Brown, and Milo Nordyke

I t is a tempting but dangerous oversimplification of the complexities surrounding U.S.–Soviet relations to think that abolishing nuclear weapons will eliminate the tensions between our two countries. It is naive to hope to escape the difficult issues posed by nuclear weapons simply by prohibiting nuclear tests. Proposed new constraints on nuclear testing involve a mixture of risks and benefits that must be evaluated in the context of overall U.S. policy. Before we can evaluate these risks and benefits, we must clearly understand the technical issues involved.

The present U.S. nuclear policy is one of deterrence, and under it, the capabilities of nuclear weapons and the ongoing nuclear test program are basic to the security of this nation. However, there is a range of ideas as to the nature of "deterrence," from existential deterrence, which asserts that deterrence can be maintained by a few survivable nuclear weapons,[1] to calculated deterrence, which relies on continued moves and countermoves by the adversaries.[2] In our view, deterrence is a dynamic condition in which we must respond to technological developments. In the Soviet Union, such developments are mainly nonnuclear and include increased air defense coverage, improved antisubmarine defenses, changing target characteristics (e.g., hardening), and increasing threats to the survivability of U.S. forces (e.g., more accurate missiles).

*This article is reprinted by permission of Science, where it appeared in the 23 October 1988 issue (vol. 238, pp. 455–460). Copyright 1988 by AAAS.

Nuclear weapon testing supports U.S. deterrence in four important ways. First, testing is done to maintain the proper functioning of the current stockpile of weapons. Second, testing is done to enhance the safety, security, and effectiveness of the existing stockpile and to respond to the changing Soviet threat. Third, testing is done to measure the effects of a nuclear weapon attack on our weapon systems and on critical command, control, and communications systems. Finally, testing is done to avoid technological surprise — to identify future weapon concepts for U.S. decision makers and to stay abreast of potential Soviet nuclear weapon developments.

One must remember that modern nuclear weapons are complex devices. Nuclear weapons produce conditions that are virtually unique — with material velocities at millions of miles per hour, under temperatures and pressures that are hotter and denser than the center of the sun, and in time scales as short as a few billionths of a second. There is no way to create these extreme conditions in the laboratory.

Nuclear warheads are designed to be enduring and robust. However, there is no such thing as a "thoroughly tested" nuclear weapon. Unlike a sampling program that tests thousands of transistors and unlike the experience gained from the continuous operation of an aircraft, a nuclear weapon typically is fully tested less than ten times during its 20-year lifetime. Any other piece of military hardware undergoes continual testing throughout its lifetime so that deficiencies can be identified and corrected. Nuclear weapons, however, are certified to function properly over a wide range of stressful conditions (e.g., temperature, humidity, shock) on the basis of a handful of nuclear tests.

Stockpile Reliability

The reliability of U.S. nuclear weapons is very high. At issue are the necessary conditions for maintaining high confidence in their reliability. Nuclear weapons are fabricated from chemically and radiologically active materials. Much as a piece of plastic becomes brittle when it's left in the sun, nuclear weapons age and change in subtle, often unpredictable ways. Some of these changes do not adversely affect their performance, but others do. Only by testing can we identify problems and determine if our solutions are successful. We know from experience that testing is essential. Fifteen of the 41 weapon designs placed in the U.S. stockpile since 1958 have required and received post-deployment nuclear tests to resolve problems.[3] In three-quarters of these cases, the problems were discovered only as a result of nuclear testing.

The provisions of the proposed treaty would be severely restrictive. A 1-kiloton yield limit would virtually eliminate our ability to maintain confidence in the nuclear stockpile or competence in nuclear technologies. Small-scale nuclear testing — below 1 kiloton — cannot today be extrapolated reliably by orders of magnitude to provide data on the functioning of a full-scale nuclear device. The fission triggers and their associated technology, which are used in U.S. strategic systems, require nuclear testing at yields greater than 1 kiloton, and partial-yield testing of the **thermonuclear (fusion)** component of most of our strategic systems must be done at yields approaching 150 kiloton. One 15-kiloton test per year would not be enough to allow us to maintain our technical skills and address all the types of problems that have arisen in the past. Even when a fix could be certified at less than 15 kilotons, we have frequently needed more than one test to fix a problem, and on occasion a problem with a particular warhead has led to concerns with other warheads.

In addition, we could not maintain scientific competence at this limited level of testing. The fundamental issue here is the quality of our scientific judgment. Nuclear weapon design is still largely an empirical science, and a designer's competence requires years of nuclear test experience. Without actual test experience, nuclear weapon scientists would lack the information needed to solve the various problems that occur with nuclear devices.

The Department of Defense and the Congress are placing more emphasis on reliability testing of radar networks, airplanes, rockets, and other military systems.[4] Nuclear weapons are more complex than any of these systems, yet the testing to ensure their reliability under all conditions is already severely limited. The proposed treaty would make matters worse. Imagine a test limit on solid rocket boosters, say, that allows partial tests of first stages, only one second-stage test per year, and no test of all three stages. Who could confidently certify the proper functioning of the rocket under these conditions? Such a test program — whether of rocket boosters or nuclear warheads — would result in a loss of reliability and confidence. It would also result in an exodus of experienced people as they left to work on other less restrictive, more productive projects.

Modernization

We believe that the general public often misinterprets the goals of the U.S. modernization program, and sees it as an attempt to perpetuate nuclear weapons. Rather, the primary focus of U.S. modernization is on the enhanced safety, security, and survivability of our nuclear deterrent forces.

As long as the U.S. has nuclear weapons, we must make them as safe and secure as possible. Although there have been no nuclear accidents involving U.S. nuclear weapons, there have been accidents in which the high explosive detonated and dispersed plutonium. Weapon designers have since devised a way to prevent this type of accident. A new insensitive high explosive (IHE) has been developed that is almost impossible to detonate accidentally, and is being used in new weapons entering the stockpile. Weapons already in the stockpile are being retrofitted with IHE, but it has been incorporated in only one-third of our systems to date. Because IHE performs much differently from previously used explosives, weapons using IHE must be redesigned and retested. Restrictive nuclear test limitations could prevent us from making this and other important changes to the stockpile.

Modern nuclear weapon safety and security features can affect the physics behavior of nuclear devices, and devices incorporating certain features can only be certified to function properly with nuclear tests. A 1-kiloton yield limit would preclude the incorporation of many safety and security measures. One 15-kiloton test per year would not provide enough test opportunities to develop new designs using IHE, for example, or to certify experimentally the yields of strategic secondaries (the thermonuclear portion of the device) mated to previously tested IHE primaries (the fission portion). Since we could not test the effects of new safety and security features on our nuclear weapons, the weapons would not be modernized with these features, without incurring large uncertainties.

Even if we use a previously tested warhead in a new system, we need a nuclear test within current yield limits (150 kiloton) to verify the new production lot. We have found through experience that we cannot specify all the detailed manufacturing criteria that affect weapon performance. Nuclear proof tests are necessary, especially when production runs last for many years and subtle changes can creep into the manufacturing process. Even an identically rebuilt warhead should be verified in a nuclear test to ensure that the slight differences from one production run to another have not affected device performance.

Weapon Effects Testing

For deterrence to succeed, our forces must not appear vulnerable to the Soviets and thus tempt them to use nuclear weapons in a crisis. We need confidence that our strategic weapons would continue to function even after a nuclear strike. We cannot know this without nuclear testing to determine the effects of nuclear weapons on components of our strategic weapon systems and on the sensors and communications equipment that would have to function after a nuclear detonation. Although above-ground nonnuclear simulators provide useful information, they do not provide a truly realistic test; they cannot, for example, provide for the synergistic effects of the various kinds of nuclear and electromagnetic radiation produced by a nuclear explosion. As in the testing of the weapons themselves, we are often surprised by the effects of nuclear testing on equipment which has performed successfully in nonnuclear tests. Equipment must then be modified and the changes certified in another nuclear test to make sure that these important elements of our deterrent will function properly.

Nuclear effects tests of U.S. equipment at 1 kiloton are not adequate today. In principle, effects tests could be performed at low yield if we moved the exposed hardware close to a 1-kiloton source. However, several problems would have to be solved — for example, damage to delicate hardware (such as a satellite) from the ground shock from the explosion. Also, we have not yet demonstrated that it is possible to develop a 1-kiloton source with a radiation spectrum characteristic of a strategic nuclear weapon.

For economic reasons, we already conduct many of our nuclear effects tests with yields near or below the proposed 15-kiloton limit, and we gain much useful information from them. However, we would be severely limited if we could field only one 15-kiloton test per year, as specified in the proposed treaty. In addition, there would be many demands for experiments other than weapon effects for that single 15-kiloton test.

Technological Surprise

If our deterrent strategy is to provide stability between the United States and the Soviet Union, we must avoid being surprised by new Soviet technology. We must anticipate changes in the threats we might face and be able to develop new systems in response to new developments. Improvements in nonnuclear features such as guidance, target hardening, and control and communications — as well as new nuclear concepts like x-ray lasing — have an impact on the effectiveness of our deterrent. We explore new weapon concepts not only with an eye to incorporating them in the U.S. stockpile but also to ensure the survivability of our forces against new Soviet threats. For all the reasons discussed earlier, these new systems will require nuclear testing.

One area where we are attempting to avoid technological surprise is the concept of a nuclear-driven directed-energy weapon (NDEW). An NDEW uses a nuclear explosive to drive a directed-energy device like a laser. At present, we are attempting to determine the viability of NDEW concepts in the hands of the Soviets to defeat a U.S. nonnuclear strategic defense system or to attack our strategic retaliatory forces in a first strike. Since we do not know how far the Soviet research has progressed, we must determine what is possible and how to defend against it. The proposed treaty would halt virtually all research on NDEW concepts. It would permit some limited research into their basic physics, but would preclude the tests that would give us an understanding of their weapon potential.

Whether any NDEW is incorporated as part of a U.S. or Soviet strategic defense system is a political decision. A very important technical question is whether Strategic Defense Inititative (SDI) systems are survivable, and nuclear testing is essential if we are to find the answer. As in nuclear-effects testing of nuclear warheads, SDI assets will have to be tested against realistic nuclear threats. Nuclear testing at current yield levels will be required until we develop the capability to perform the necessary tests at lower yields.

Computer Simulations and Nonnuclear Testing

Critics of nuclear testing have frequently asserted that a viable nuclear deterrent could be maintained with nonnuclear and very low-yield nuclear testing plus computer simulations. A variation of this argument is that although new warheads could not be developed with such testing and computer simulations, they would be adequate for maintaining a stockpile of existing weapons. Neither of these assertions is valid.

The problem lies with the unique nature of nuclear explosives. A nuclear explosive involves myriad physical processes — from the macroscopic down to the microscopic — and they are all interrelated. A nuclear explosion involves most of the physics of a supernova, and the academic community has been working on a computational description of these processes ever since computers were developed. A nuclear explosion also is affected by the microscopic detail of engineering and materials (assembly gaps and grain structure). It simply is not possible with today's computers and computing techniques to include the full range of processes and level of detail in a simulation.

In a computer simulation of a nuclear explosion, we attempt to provide a detailed physics model for all of the interrelated, nonlinear processes that occur. However, computer simulations are inherently limited because (1) the physics must be approximated by numerical algorithms, and these approximations are of varying degrees of accuracy; (2) not all of the physical processes can be included in detail, given the physical limitations of the computer facilities; and (3) experimental data are rarely available to confirm the appropriateness of the level of detail in the simulation. In addition, many of the phenomena are interrelated, and so errors from a simulation of early processes will propagate through simulations of subsequent processes. Thus, a small error in an early step can grow to yield a calculated result that bears little, if any, resemblance to the results of an actual test. Usually but not always, our simulations correctly predict general trends in device performance, but sometimes correct detail and important performance parameters can elude us completely.

To minimize potential errors, we normalize our calculations — to the extent we can — to the results from actual experiments. Although usually we can recognize that there is an error in a simulation, it is very difficult to identify specifically what is wrong because of the paucity of actual data. The conditions that occur in a nuclear explosive are so unique that we can obtain valid data only from a nuclear explosion, difficult as it is to conduct experiments in so harsh an environment. No experimental facility other than a nuclear explosion itself can give us data about what actually happens in a nuclear explosion. Detailed information accumulates slowly because we field only a limited number of nuclear tests each year.

We make extensive nonnuclear tests on those parts of the system that are amenable to such tests (the high explosive and electrical systems, for example). We then attempt to extrapolate these results to the energy regime of a nuclear explosive (many orders of magnitude greater). Unfortunately, we find that the data from nonnuclear testing, coupled with our most sophisticated calculational procedures, cannot be extrapolated to accurately predict the behavior of a nuclear device.

This problem with extrapolating the results of small-scale tests and computer simulations is not unique to nuclear devices. It is also the case in modern nonnuclear weapons. For example, in modern rockets, small-scale tests and computer simulations do not accurately predict the detailed behavior of solid rocket propellant. Some full-scale tests and actual launches are needed to certify the rocket's proper functioning.

Nuclear tests are particularly important for boosted primaries. Boosting is a process that greatly increases the yield obtained from the fission primary and makes it possible to use much smaller primaries in modern strategic nuclear weapons. However, the boost process is complicated and not fully understood, and some of the stockpile problems encountered to date have involved concerns about inadequate primary boosting. If boosting of the primary is less than expected, proper ignition of the secondary may not occur, and it will fail to produce its designed yield. Hence, to certify the proper functioning of a warhead with a boosted primary, we must be able to certify proper boosting and ignition of the secondary. This requires nuclear testing at yields greater than 1 kiloton; we cannot today reliably extrapolate the results of a subkiloton test to the performance of a full-scale primary.

A recent example illustrates the essential role of nuclear testing to verify boosting. Just as a new weapon was being deployed to the stockpile, we made a final proof test at the weapon's specified low-temperature extreme. The test results were a surprise. The primary gave only a small fraction of its expected yield, and this was insufficient to drive the secondary. The weapon had been tested extensively in nonnuclear hydrodynamic tests, even at its low-temperature extreme generating no serious concern. On the basis of the nonnuclear testing, previous successful nuclear tests, and extensive computer modeling, we had every reason to expect that the low-temperature proof test would produce the predicted yield. However, this low-temperature nuclear test revealed that something was not right. After extensive post-test analysis, the problem was identified and the design was modified. Another low-temperature nuclear test was performed and this test was successful, establishing confidence that the warhead would operate properly over its entire temperature range. The production specifications were changed, and the approved, modified warhead entered the stockpile. At present, this stockpile is extremely reliable. But it is reliable only because continued nuclear testing at adequate yields allows us to identify and correct problems as they occur.

The Impact of Restrictive Test Limits on the Soviets

We can only surmise the effects of further test limits on the Soviet Union. Since 1963, when the Soviet Union and the United States agreed to conduct nuclear tests only underground, we have learned little about the Soviet nuclear weapon program. What we do know indicates that the Soviets have an aggressive, well-funded program with impressive technical achievements. We know from their unclassified literature that they understand the physics of x-ray lasers. Since nuclear weapon technology is not monolithic, the Soviet designs could be very different from ours. On the basis of the Chernobyl reactor accident, one could infer that they have a different attitude about the enhanced safety and security features that add complexity to U.S. warheads. We also know that the Soviet missiles have a very large throw weight. They could use this large throw weight to accommodate warheads that are less technologically sophisticated (and thus larger and heavier) than U.S. designs. The U.S. approach of incorporating sophisticated technologies in its nuclear warheads and in virtually all its military equipment has many important benefits, but it also has attendant costs, including a greater reliance on testing to ensure proper functioning.

We believe that the impact of the proposed treaty's limits would be less harmful to the Soviets than to the U.S. Given their apparent reliance on larger, less sophisticated, and less complex weapons, deterioration of Soviet systems would likely be less of a concern. Restrictive state policies could ensure the retention of their scientific base. Their closed society could allow them to exploit shortcomings in verification at 1 and 15 kiloton. It would also enable them to secretly prepare for a treaty breakout, as they did before, during the nuclear test moratorium of 1958–1961.

Verification of the Proposed Treaty

Measures to verify compliance with a treaty must enable us to detect any militarily significant clandestine tests, with sufficient certainty. If uncertainties are too high, we have a situation like that with the Threshold Test Ban Treaty (TTBT), where our yield estimates are not sufficiently precise to permit us to make definitive judgments about Soviet compliance with the 150-kiloton limit. This inability to distinguish reliably between compliance and potential violations can undermine confidence in the arms control process and can heighten international tensions. Therefore, before we enter into new treaty obligations, we must distinguish between proven verification techniques and possible future capabilities.

The proposed treaty presents a number of verification issues that must be clearly defined and analyzed. Before we can negotiate any treaty involving such a technically complex verification regime, myriad details — many of them not yet apparent — must also be clearly defined. Because of the absence of a comprehensive verification proposal, we will only address the general verification issues raised by the proposed treaty.

A treaty permitting tests with yields up to 1 kiloton on declared test sites and prohibiting all tests outside those sites would require two separate verification regimes. The first would measure the yields of the permitted tests at the test sites to verify compliance with the 1-kiloton limit. The second would verify that no clandestine tests were conducted anywhere else in the country; this would be very similar to the regime required to verify a Comprehensive Test Ban Treaty (CTBT). The proposed treaty introduces two other verification requirements — measuring the yield of one test each year with a yield up to 15 kiloton, and determining that this test consists of only one nuclear device.

Let us first address the country-wide, CTBT–type regime required to preclude clandestine tests outside the test sites. The primary evasion modes are (i) exploding the device in a large underground cavity to decouple the shock wave from the earth and thus reduce its seismic signals; (ii) hiding the seismic signal in the coda (that is, the final portion of the seismic signal) of a large earthquake or chemical explosion; and (iii) conducting the explosion in outer space. It should be made clear that the problem of detecting such evasive tests is greatly complicated by the background of natural or licit events that give rise to false alarms that must be discriminated against.

We would need 25 to 30 high-quality, in-country arrays, or equivalent single stations with high signal-to-noise ratios, located inside the Soviet Union to detect signals from kiloton-sized, cavity-decoupled nuclear explosions with high confidence.[5] Such a network would also greatly improve our ability to detect clandestine nuclear explosions with yields of 5 to 10 kiloton or larger hidden in the coda of a large earthquake.

However, detection does not constitute identification. There are thousands of earthquakes each year in the Soviet Union with magnitudes comparable to decoupled, kiloton-scale nuclear explosions. The magnitude level at which we have 90 percent confidence for identification is presently three to four times greater than the magnitude level for detection. Many seismic events are detected that cannot be identified. There are also hundreds of chemical explosions each year that have seismic signals in this same range and thus cannot be discriminated from nuclear explosions. Thus, it is obvious that there will be many unidentified seismic events each year that could be decoupled nuclear explosions with militarily significant yields much greater than 1 kiloton. On-site inspections (OSI) have been suggested as a means for resolving these concerns. However, it is not clear what such OSIs would consist of, how many would be allowed, and what we could expect them realistically to accomplish.

It has been argued that high-frequency seismic signals can be used to significantly improve detection and discrimination of earthquakes from partially or fully decoupled nuclear explosions.[6] This argument is based on downward extrapolations of results from seismic events that are much larger than those relevant to a CTBT or a 1-kiloton limit. This argument also makes several assumptions about the high-frequency propagation properties of the upper mantle and crust in the Soviet Union. High-frequency seismic signals may well be useful, but their utility and reliability for detection and identification must be demonstrated in many monitoring environments before we can fully assess their contribution to treaty verification. We are actively working to improve our understanding of these difficult issues.

The testing of nuclear devices in deep space is also a serious concern. Existing rockets could be used to launch a deep-space mission containing a nuclear device and a diagnostics package. All earth-orbiting nuclear-test detection sensors have a detection range that is limited by background noise levels. At a distance well beyond the detection limit of earth-orbiting sensors (determined by the planned yield), the two packages would be separated, the nuclear device fired, and the performance data collected from the diagnostics package and sent back to earth. The time required for such a mission would vary from a month to a year, depending on the yield. It would be much simpler to field than most deep-space probes because the total life of the mission is limited, no external power sources are required, and the experimental measurements are relatively simple, highly automated, and well proven. The only practical way to prevent such tests would be to inspect all space flights large enough for such a mission before launch and to search for highly enriched fissile material.

Proponents of the proposed treaty assert that the ability to conduct one 15-kiloton test each year would eliminate the motivation to field clandestine tests of this magnitude. However, if one country carried out even two or three more tests each year at 5 to 20 kiloton than the other country, it would gain a substantial advantage in nuclear weapon technology.

Under a treaty that permits unlimited numbers of 1-kiloton tests and one 15-kiloton test per year at designated sites, the primary task would be to verify that the yields are within the allowed limits. Provisions have been suggested to restrict such tests to one test area in hard rock. These provisions attempt to minimize the uncertainties of estimating yields seismically by ensuring that the explosions are well coupled.

However, U.S. experience with estimating the yields of low-yield tests from seismic signals shows that the results are quite variable and easily affected by local geologic structure. Tests in hard, competent rock overlying weak, porous layers can appear to have significantly smaller yields than is actually the case. A country could exploit such a geologic situation to field tests at several times the threshold, thereby gaining a militarily significant advantage. Such problems in unknown or poorly documented geological environments could, in theory, be minimized by requiring all tests to be conducted below the water table. Unfortunately, the U.S. has virtually no experience with low-yield tests under these conditions.

Calibration explosions, with yields determined by measuring the speed of the hydrodynamic shock wave, have been suggested[7] as a way to reduce the uncertainties associated with seismic coupling and transmission. However, there are many operational and geometrical problems with this technique at low yields, and its accuracy has not been established in this yield range.

Another suggestion has been to use multiple seismic phases to reduce the uncertainty associated with seismic yields.[8] The validity of this approach is being reviewed. This technique, should it prove acceptable for explosions in the 15-kiloton range and below, would require data from stations inside each country to detect the required regional phases. We would also have to calibrate these stations with a significant number of calibration explosions at the designated test sites.

Because of space limitations, our discussion here is based on many assumptions about provisions of access to each country's territory. The details of such verification provisions are crucial to a verification agreement. Even if such details could be worked out, the shortcomings in our knowledge and capabilities limit our ability to ensure compliance with the provisions of the proposed treaty.

Conclusions

Nuclear weapons that are safe and secure, reliable, survivable, and effective will be a critical element of this nation's deterrent for the foreseeable future. The existence of these weapons reflects the tension that exists between the United States and the Soviet Union. Nuclear test bans will not reduce or eliminate nuclear weapons or this tension. Imprudent nuclear test bans, however, could impair the viability of this vital element of U.S. security.

New, more restrictive test limitations would not enhance our national security. They do not address the two most important issues — namely, major reductions in strategic and conventional forces of both the Soviet Union and the U.S., and a widespread lessening of tension between our two countries. In fact, it is conceivable that the diversion of political attention from arms reduction efforts and the distrust generated by test ban verification problems could actually increase tensions between the two countries.

We believe that more restrictive test limitations or a nuclear test ban should be considered only as part of an integrated and comprehensive approach to arms control. We must reduce the numbers of the most destabilizing weapons and the overall size of the strategic arsenals through negotiations. A restrictive test ban may be a proper last step in our quest for nuclear arms control and a stable peace, but it would, in our opinion, be an imprudent first step. Further test limitations will be consistent with increased stability and decreased tension between the U.S. and Soviet Union only if they are instituted after major stabilizing reductions are made in the strategic nuclear and conventional forces of both countries.

Notes

1. Robert S. McNamara, *Blundering Into Disaster, Surviving the First Century of the Nuclear Age* (Pantheon Books, New York), 1986.

2. See, for example, Secretary of Defense Harold Brown's Posture Statements, *Department of Defense Annual Report, FY 1979 and FY 1980.*

3. P. S. Brown, "Nuclear Weapons R&D and the Role of Nuclear Testing," *Energy and Technology Review*, Lawrence Livermore National Laboratory, Livermore, CA, September 1986.

4. Newsletter, Rep. Fortney P. Stark (California), August 1986; also, Biddle, W., "How Much Bang for the Buck," *Discover*, September 1986.

5. W. J. Hannon, "Seismic Verification of a Comprehensive Test Ban," *Science* 227, pp. 251–257 (January 18, 1985).

6. J. F. Evernden, C. B. Archambeau, and E. Cranswick, "Evaluation of Seismic Decoupling and Underground Nuclear Test Monitoring Using High Frequency Seismic Data," *Reviews of Geophysics*, Vol. 24, No. 2, pp. 143–215, May 1986.

7 U.S. Policy Regarding Limitations on Nuclear Testing, U.S. Department of State Special Report No. 150, August 1986.

8. P. Richards, Testimony before the Senate Committee on Armed Services, February 26, 1987.

23

White House Space Policy: How It Is Made and Its Rationale*

Roger G. DeKok

W hen I first arrived in my current position, I faced a wide variety of opinions. Many people argued that it was time for a change in our national space policy. Others have said there is no space policy. Still others alleged that our space programs lack leadership.

I would like to dispel some misconceptions. National space policy is not something developed by a single person or a very few people who then impose it on a submissive federal bureaucracy. Nor is it a collection of programmatic decisions that circumvent the normal budget process. Rather, national space policy establishes the fundamental principles, goals and strategy that guide U.S. government space efforts. It is developed within an interagency process and is approved by the president.

I am frequently asked why a revised national space policy needed to be issued so "late" in the course of President Reagan's second term. In response, I cite four reasons:

- It has been over five years since the last comprehensive space policy was issued.

- Some unexpected events have occurred, in particular the Shuttle and expendable launch vehicle failures that plagued our space program in 1985

*This article is condensed and updated from a speech given at a luncheon for the Washington, DC,. Section of the American Astronautical Society, February 18, 1988, that also appeared in the May/June 1988 issue of The Space Times.

and 1986, and the corrective actions that were necessary in the aftermath of these events.

- Presidential decisions to foster the development of a commercial space industry needed review to determine if further guidance to advance these policies was necessary. Other presidential decisions, specifically those to build a permanently manned space station and to begin the Strategic Defense Initiative, were made after the 1982 policy was issued. These items were integrated into the new policy.

- Continued growth in foreign space programs, both civil and military, warranted an examination of America's relative position.

The review of national space policy was conducted against a backdrop of criticism that U.S. space capabilities — for years, indisputably the world's finest — have been eclipsed by the growing space exploits of the Soviets, and are increasingly being challenged by those of the European Space Agency and Japan.

For those reasons, a review of national space policy was begun within the existing interagency process established by the president to periodically review space issues: the Senior Interagency Group for Space, or as it is commonly called SIG(SPACE). SIG(SPACE) is chaired within the National Security Council, and is comprised of representatives from the National Aeronautical and Space Agency (NASA); the Departments of State, Defense, Commerce, and Transportation; the Central Intelligence Agency (CIA); the Office of the Joint Chiefs of Staff; the Arms Control and Disarmament Agency; the Office of Science and Technology Policy; and the Office of Management and Budget. In addition, the Treasury Department participated in this policy review.

In reviewing and revising the national space policy, SIG(SPACE) was assisted by its lower-level Interagency Group for Space, IG(SPACE), and some working groups chartered by IG(SPACE) to develop policy recommendations in several areas. These policy deliberations were extensive.

For example, the prime IG(SPACE) working group met a total of 22 times for a total duration of over 100 hours to produce a document that is less than 20 pages long. No paragraph, sentence, or even word was placed in the final document without considerable debate. The process began on the last day of July 1987, and the SIG(SPACE) deliberations were finished in mid-December. Administrative details and the preparation of the decision package for the president ensued, leading to presidential approval on January 5, 1988. The final policy is documented in the form of a National Security Decision Directive, which is classified. An unclassified version of this policy was released by the White House Press Office on February 11, 1988.

In producing the revised policy, SIG(SPACE) representatives used a wide variety of source documentation, including previous presidential decisions, information and proposals developed by the Economic Policy Council's Commercial Space Working Group, the National Commission on Space (Paine Report), Dr.

Sally Ride's report on "Leadership and America's Future in Space," testimony before congressional committees, as well as numerous writings available on the subject within the public domain.

A major source of data, especially on the topic of leadership, was comparisons of U.S. space activities versus those of other countries. The summary conclusions of these comparisons were as follows:

In the national security area, U.S. space capabilities, under conditions short of direct attack on our space systems, are clearly superior to those of our potential adversaries. As former Defense Secretary Casper Weinberger stated several months ago, "In terms of operational military capability, now and in the future, the U.S. exceeds equivalent Soviet capability in terms of the quality, accuracy, and timeliness of mission data to the users." (He might have added the longevity of orbiting systems, as well.)

However, in the event of a conflict involving attacks on space systems, our technological lead would tend to be offset by demonstrated Soviet antisatellite capabilities for which the U.S. has no counterpart.

Nonetheless, the national security space sector has taken a number of steps to ensure continued mission capability, even if we experience failures in our on-orbit or launch assets, whether from natural causes or hostile action.

The Soviets possess an extremely robust and diverse stable of space launch capabilities that appears to be far in excess of what they need, given our current understanding of their launch requirements. This diversity, which the U.S. is now striving for in the aftermath of the Shuttle accident, provides the Soviets with some inherent advantages. To offset these advantages, the U.S. needs to complete its launch recovery program, field an antisatellite system of its own, and continue to improve space systems survivability.

In the civil area, the assessment revealed that the space capabilities of our competitors are indeed growing and, in some cases, at a more rapid rate than ours. However, in critical areas, U.S. technological capabilities remain the best in the world.

It is clear, though, that the launch hiatus has diminished the traditional U.S. lead in several key science and exploration areas. This is a trend that will continue until the Shuttle is returned to safe, reliable operation and we begin to launch the backlog of important civil payloads that are awaiting access to space. Apparently unique Soviet requirements have prompted resource commitments in manned spaceflight and space transportation systems that far exceed those of the U.S. These commitments leave the Soviets well positioned to translate infrastructure potential into real scientific accomplishments over the next few decades.

In commercial space systems, U.S. efforts, although still embryonic, promise important economic, industrial, and national security benefits as long as government policies continue to provide a climate conducive to sustained growth.

As a direct result of these policies, American firms are aggressively marketing launch services worldwide and, to date, U.S. companies have signed contracts to launch 18 satellites with expendable launch vehicles, contributing approximately $660 million to the U.S. balance of trade. Investments totaling approximately $400 million have been made in this emerging business by commercial companies, which may result in the creation of some 8,000 new jobs.

The new national space policy is organized into two parts: a "policy" section that provides a synopsis, and a longer section of "policy guidelines and implementing actions" that provides a more detailed framework through which the policies are to be carried out. The highlights of this new policy are:

- Space leadership is recognized as a fundamental national objective guiding U.S. space activities. However, the directive states that "Leadership in an increasingly competitive international environment does not require United States preeminence in all areas and disciplines of space enterprise. It does require U.S. preeminence in key areas of space activity critical to achieving our national security, scientific, technical, economic, and foreign policy goals."

- "Leadership" had been cited previously as a fundamental goal of U.S. space activities; unfortunately, it was not further defined or qualified, and, as a result, was often cited as the reason for doing whatever any particular advocate wanted to do in space.

The new policy recognizes that "leadership" cannot and should not be universal, nor can it be merely proclaimed. Rather, leadership consists of establishing important goals and priorities and acquiring the means to achieve them. In that regard, the policy states the overall goals of U.S. space activities as:

- Strengthening the security of the United States;

- Obtaining scientific, technological, and economic benefits for the general population, and to improve the quality of life on earth through space-related activities;

- Encouraging continued United States private sector investment in space;

- Promoting international cooperative activities, taking into account U.S. national security, foreign policy, and scientific and economic interests;

- Cooperating with other nations in maintaining freedom of space for all activities that enhance the security and welfare of mankind; and

- A long-range goal, expanding human presence and activity beyond Earth orbit into the solar system.

This last goal is new and has received a great deal of publicity, frequently of the misleading variety. This goal establishes the general direction and focus for efforts and technologies guiding the nation's civil space sector. It is not a commitment to any particular mission, destination, or timetable. It is a commitment to a broad-based technology program called "Pathfinder," directed by NASA, to examine the challenges associated with manned spaceflight into our solar system. Pathfinder is aimed at acquiring the know-how to allow wise future decisions leading toward more focused programs for manned exploration.

Some will applaud the vision embodied in this goal. Others may lament that it does not go far enough. We believe that we must learn more about the long-term aspects of living and working in space — identifying and meeting the technical challenges and costs — before deciding when manned planetary missions might be appropriate. To do otherwise, by committing prematurely, for example, to a manned mission to Mars by a certain date, could turn out to be a hasty, costly, and even dangerous decision, based on current data and technology.

Many accounts have speculated that this policy will result in a near-term decision to cooperate with the Soviets in a future manned mission to Mars. While international cooperation is a goal of U.S. space policy, it is premature to speculate on such cooperation. The U.S. has not committed itself to a manned Mars mission, and the current budget situation makes such an outlook, even in the future, difficult.

We are presently rebuilding our space cooperation relationship with the Soviet Union after a five-year interruption, and it will take some time to restore confidence to the level at which more ambitious cooperative projects could be considered. Indicative of the importance attached to the international cooperation, President Reagan and General Secretary Gorbachev agreed at the recent Moscow Summit to several initiatives to expand the 1987 U.S.–Soviet space cooperation agreement. They agreed to exchange flight opportunities for scientific instruments to fly on each other's spacecraft, and to exchange the results of independent national studies of future unmanned solar systems exploration missions as a means of assessing the prospects for further U.S.–Soviet cooperation on such missions. They also noted scientific missions to the Moon and Mars as areas of possible bilateral and international cooperation.

There are always the troublesome (but often solvable) aspects of technology transfer that arise when highly sophisticated technical projects of this nature are contemplated. The near-term focus in international cooperation associated with manned spaceflight continues to be NASA's Space Station program, emphasizing the cooperation with friends and allies that the president is seeking.

In June 1988, an important milestone in this program (which the president has named "Freedom") was achieved with NASA's announcement that negotiations among the U.S., Canada, Europe, and Japan on the framework for international cooperation in the Space Station program have been completed. Spanning decades, Freedom will be the largest international scientific and technological venture ever undertaken.

The new goal of human expansion does not mean that space science is being downgraded or that the NASA program of unmanned space exploration is being abandoned. Quite the contrary. Under this policy, the first objective of U.S. civil space activities is to expand knowledge of the Earth, its environment, the solar system, and the universe. The policy reaffirms the long-standing national objective of supporting a vigorous and far-reaching program of space science. The policy guidelines state that NASA will conduct a balanced program of manned and unmanned exploration.

We need both manned and unmanned exploration, with determinations made on the basis of cost, safety, suitability, and expected results, given the specific mission objectives involved.

In this policy, for the first time, commercial space activities are recognized as a separate and coequal element of U.S. space endeavor. In furtherance of other administration policies, this space policy encourages private sector space enterprise by establishing guidelines to ensure that Government agencies will become reliable customers for space goods and services; limiting Government competition with commercial space enterprise; and providing that Government regulation of this industry be limited in extent to that required by law, national security, international obligations, and public safety.

In the national security sector, a policy of assured access to space is recognized as a key element of National Space Policy for all space sectors. U.S. space transportation systems must provide a balanced, robust, and flexible capability with sufficient resiliency to allow continued operations despite failures in any single system. In particular, payloads will be distributed among launch systems and launch sites to minimize the impact of loss of any single launch system or launch site on mission performance.

The policy reaffirms the prohibition against NASA maintaining an expendable launch vehicle (ELV) adjunct to the Shuttle, as well as the limitation on commercial and foreign payloads on the Shuttle to those that are Shuttle-unique or serve national security or foreign policy purposes. Policies endorsing the purchase of commercial launch services by government agencies, instead of government procurement of the vehicles, are further strengthened. These policies should provide additional impetus to our developing a domestic U.S. commercial launch industry.

Finally, policies for civil earth remote sensing have been established to encourage the development of U.S. commercial systems competitive with or superior to foreign-operated civil or commercial systems. To implement this policy, guidelines are established stating that there are no predetermined limitations or restrictions on the performance of civil earth remote sensing systems. In licensing these systems, the federal government will consider national security and foreign policy factors, including those required by law. Such considerations have not precluded licensing in the past.

The new policy also contains provisions for controlling the growing space debris problem, continuing government research and development (R&D) support for advanced space communications technologies, and investigating financial alternatives for future space infrastructure developments.

In summary, the new national space policy provides a sound structure for guiding U.S. space activities far into the future. It balances the goals of governmental space sectors, recognizes the potential of the nongovernmental commercial space sector, adopts a new goal that will better focus NASA's long-range plans, and establishes the framework for U.S. cooperative space efforts.

The message is clear: America's future is tied to its capabilities and aspirations in space. Moreover, its commitment to space leadership has been reaffirmed, and no one should question the strength of this commitment. We are prepared for a bright future. Once again, America's technology, imagination, and hope will converge to enable the conquest of space.